Limeng Wang

# Image Analysis and Evaluation of Cylinder Bore Surfaces in Micrographs

**Forschungsberichte aus der Industriellen Informationstechnik
Band 9**

Institut für Industrielle Informationstechnik
Karlsruher Institut für Technologie
Hrsg.  Prof. Dr.-Ing. Fernando Puente León
       Prof. Dr.-Ing. habil. Klaus Dostert

Eine Übersicht über alle bisher in dieser Schriftenreihe erschienenen Bände
finden Sie am Ende des Buchs.

# Image Analysis and Evaluation of Cylinder Bore Surfaces in Micrographs

by
Limeng Wang

Dissertation, Karlsruher Institut für Technologie (KIT)
Fakultät für Elektrotechnik und Informationstechnik
Tag der mündlichen Prüfung: 24. September 2013
Referenten: Prof. Dr.-Ing. F. Puente León, Prof. Dr.-Ing. T. Längle

**Impressum**

Scientific
Publishing

Karlsruher Institut für Technologie (KIT)
KIT Scientific Publishing
Straße am Forum 2
D-76131 Karlsruhe

KIT Scientific Publishing is a registered trademark of Karlsruhe
Institute of Technology. Reprint using the book cover is not allowed.

www.ksp.kit.edu





Print on Demand 2014

ISSN  2190-6629
ISBN  978-3-7315-0239-5
DOI   10.5445/KSP/1000041878

# Image Analysis and Evaluation of Cylinder Bore Surfaces in Micrographs

Zur Erlangung des akademischen Grades eines

## DOKTOR-INGENIEURS

von der Fakultät für

Elektrotechnik und Informationstechnik

des Karlsruher Instituts für Technologie (KIT)

genehmigte

## DISSERTATION

von

**M.Sc. Limeng Wang**

geb. in Luoyang, China

Tag der mündl. Prüfung:   24.09.2013

Hauptreferent:   Prof. Dr.-Ing. F. Puente León, KIT

Korreferent:   Prof. Dr.-Ing. T. Längle, Fraunhofer IOSB

# Acknowledgement

The completion of my dissertation and subsequent Dr.-Ing. has been a long journey. During this process I have experienced confusion, hesitation and depression. The success of my doctoral study would not be achieved without the support, the help and the encouragement of the people by my side. Except the extended knowledge I also acquired the valuable life experience during the stay at the institute, and in Germany. That will greatly benefit my future life. Therefore, I would like to gratefully and sincerely express my gratitude.

First of all, I extremely appreciate my mentor, Prof. Fernando Puente León. He provided me this research thesis, when I was still in China. His valuable guidance and scholarly inputs opened my mind and directed me to the correct road at various phases of this research. He has always made himself available to clarify my doubts despite his busy schedules. I thank him for the unconditional support and the facilities provided to carry out the research work at the institute. Special thanks also to Prof. Thomas Längle, who was willing to review my dissertation as the co-referent. His concise comments has great help for the revising of this paper.

Everything is hard at the beginning. It is true not only for the research but also for the integration into the new environment. Luckily, I received generous kind help from the secretary and other working colleagues at the institute. My gratitude is hereby extended to Mrs. Koffler. Since my enrollment at the faculty and up until my doctoral examination, she has always contributed to the organization issues for me. I owe also a very important debt to Dr. Konrad Christ, Dr. Andreas Sandmair, Matthias Michelsburg, Mario Lietz and others, who did great effort for the organization of institute events. In particular, I acknowledge Dr. Wenqing Liu, for being the discussant for the feeling and experience during the doctoral study. I am very grateful for the convenient working atmosphere at the institute.

I would like to offer my special thanks to Deutsche Akademische Austauschdienst (DAAD) for the scholarship, which was used to finance my living cost during the doctoral study. I owe a lot to Prof. Klaus Dostert for his recommendation to prolong the scholarship every year. I will also remember the life at Chinesisch-deusches Hochschulkolleg (CDHK) of the Tongji-University in Shanghai, where I gained the master degree as well as the chance of the doctoral study in Germany. Prof. Elmar Schrüfer of the Technical University of Munich is a warm-hearted tutor for each overseas student of CDHK. Without his encouragement, this dissertation would not have materialized.

Of course, no acknowledgement would be complete without giving thanks to my parents. They have taught me about hard work and self-respect, about persistence and about how to be independent. I must give the tremendous and deep appreciation to my wife, Qingqing. Through her love, patience, support and unwavering belief in me, I have been able to complete this dissertation journey.


Lenting, in May 2014                                    Limeng Wang

# Contents

# Contents

# 1 Introduction

Internal combustion engines are the heart of modern automotive vehicles. The cylinder crankcase, as the core component of the engine, provides an enclosed space for generating the vehicle power. Cylinder liners are like the ventricles, in which pistons send rhythmic heartbeats to drive the engine. For a traveling car the rotation speed of a motor is normally more than 1000 r/min . This means that the friction occurs quite intensively between the piston and the cylinder wall. Surface structures of the cylinder wall have important influence on friction losses, oil consumption and piston wear. These factors further influence engine performances like energy efficiency, noxious emission and service longevity. For these reasons the quality of cylinder bore surfaces is of great interest in the field of engine production. Inspection techniques based on image analysis have been commonly used for the quality control of engine cylinders. Inspired by the increasing demand on inspection robustness and precision, this thesis focuses on novel inspection solutions for cylinder bore surfaces that possess a wide range of qualities.

## 1.1 Background

Surface textures on cylinder bores play an essential role for reducing friction and oil consumption. Previous motor tests manifest that tribological processes in the cylinder-piston system, as shown in Fig. 1.1, are very complex. Several different theories [13, 64, 66] have been presented to characterize the relations between the surface topology and engine performances. Based on these theories, a variety of manufacturing methods were developed to finish function-relevant cylinder liner surfaces. The commonly used production approach is the honing technique, because it can significantly improve the geometrical accuracy of cylinder bores and create specially designed surface structures. Engines finished by honing are desired to fulfill the increasing legislative requirement for harmful emission, as well as to satisfy customers' demands on environment-friendly and energy-efficient automobiles. For these purposes, image-

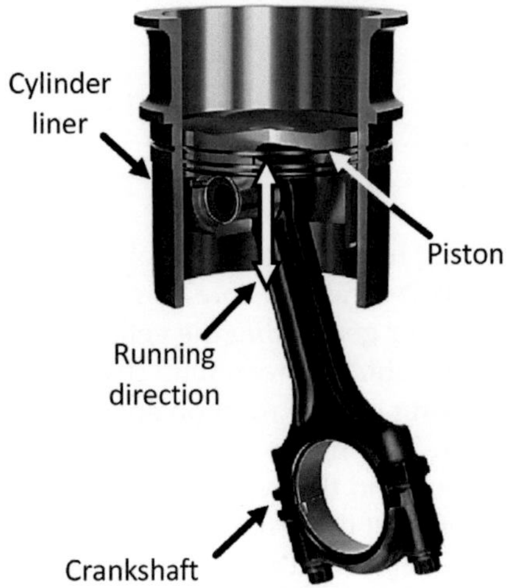



**Figure 1.1**  Cylinder–piston system in automotive engines.

based surface inspection is widely used in the research of manufacturing techniques as well as in the quality assurance for mass production.

The evaluation of new honing techniques is normally performed offline in laboratories. It has been known that triboogical properties of honed surfaces can be improved by optimazing the cylinder material, the honing tool or the honing process. Honed cylinder liners are carried into the laboratory. Then, quality features describing textural and topological surface structures are measured in a manual or an automated fashion. In this process, it is verified whether honed surfaces conform to the design specifications. The quality control for the mass production requires the same measurement of surface features as that conducted in offline applications. Surface qualities are evaluated according to industrial standards that are formulated by car manufacturers. However, not all inspection tasks are online realisable. A part of inspection tasks concern material details, such as metal folds and graphite grains. This requires highly resolved surface data. Some 2D and 3D imaging systems are commonly used by car manufacturers. For example, light optical microscopy and scanning electron microscopy are adopted to investigate 2D texture features and material structures. Due to the complex internal structure of microscopic systems, it is still difficult to integrate them in machines. Besides, white light interferometers and the confocal microscopy are useful measuring instruments for roughness measurement. With 3D imaging systems, honed surfaces are described by so-called 2.5D images which encode surface heights. Image analysis methods can thus be used to analyze surface textures as well as surface roughness. Since 3D imaging systems are very sensitive to the mechanical vibration, currently they are not suitable for online applicaitons. Nevertheless, several researches [36, 52] were conducted to design non-distructive 3D inspection systems. These systems still cannot be integrated in production lines. Besides, 2.5D images are acquired by scanning the surface topology. Thus, the inspection process is time-consuming. To facilitate the acquisition of surface data, small surface specimens are extracted at different places of cylinder walls by splitting cylinder liners. With the aforementioned imaging systems, surface details with a size of micrometers can be clearly revealed in images. As explained before, such a high image resolution is still diffclut to realise in online inspection systems. Other inspection tasks concern the overview evaluation of machining patterns (also known as surface finishes), such as the measurement of honing angles, the balance of honing grooves as well as the recognition of blowholes, bubbles and scratches with a size of millimeters. These quality problems can be investigated in macro-scales. Macroscopic imaging systems are already good enough for these tasks. This enables fully automated 100%-inspection of cylinder bore surfaces. Some compact 2D imaging systems [28, 45, 48] insert images sensors into cylinder bores and scan cylinder walls around $360°$. With these systems, images showing complete cylinder walls can be directly acquired in cylinder bores. The inspection process is fast and non-distructive. These systems are suitable for both on- and offline appliactions. Compared with the conventional visual inspection, results generated by image analysis are reliable and reproducible. Hence, automated vision systems are favored in the automobile industry. In the meantime higher requirements are put forward to the performance of inspection systems.

## 1.2 Motivation

This thesis develops image analysis algorithms for standard 2D microscopic systems. The highlights of the work lie in the algorithm design for two challenging inspection tasks — the detection of metal folds and graphite grains. By assuming that the micrographs of honed surfaces show a good image quality, the research aims to overcome the difficulties for analyzing complex surface details.

In the field of automated inspection of honed surfaces, the research interest lies in three apsects — image acquisition, image processing algorithm, and quality parameters. This thesis contributes to the last two aspects. The work does not actually improve imaging systems. However, the choice of proper methods for image acquisition builds the fundation of the work. The tendency in image acquisition techniques is to non-destructively capture the surface data directly in the cylinder bores. As long as surface images can "faithfully" reflect the surface topography, the proposed image processing algorithms can be commonly applicable to both destructive and non-destructive imaging systems. In this thesis, a destructive way to capture surface images is adopted. Surface samples are prepared by splitting cylinder bores. Then, two standard approaches for image acquisition are exploited. Pictures captured by a light optical microscope (LOM) and a scanning electron microscope (SEM) are adopted for investigating the surface appearance at different scales. The choice of imaging systems hat a reason that surface components influence engine properties in their own roughness scales. An inspection method that handles honed surfaces at a certain observation scale may lower the accuracy of describing relevant objects at other scales. Hence, a more rational strategy for surface description is to match the image acquisition method with the scales of surface components.

This thesis focus on the further development of inspection methods for 2D surface finishes. Two novel algorithms for defect inspection in SEM images and graphite detection in LOM images are proposed. In the first algorithm a local approach for recognizing surface defects overcomes the drawbacks of conventional model-based methods. It can be noted that tool marks imaged in micrographs may be far from the ideal model of honing textures. The image analysis for these surface objects is more difficult than that for regular honing patterns. In previous works [2, 14, 80], honing grooves were assumed to dominate the surface. Thereby, texture features could be extracted from the entire image. Defects were detected as the abnormal of honing textures. However, in the worst case, expected honing textures may be seriously degraded in the surface finish due to an improper manufacturing process. The model-based analysis of honed surfaces may lead to instable evaluation because of poor surface qualities. Therefore, more effective and robust surface features are desired to deal with a variety of honing structures. The presented approach for defect inspection is edge-based. Edge features like the orientation and the strength are used for constructing feature maps. To ensure the robustness of feature extraction, a novel orientation estimator is designed. The presented inspection strategy shows some significant advantages:

- Independent of the formation of honing textures,

- Applicable to a wide range of surface qualities and

- Potential to be extended to 2.5D applications.

The second algorithm concerns the investigation of laser-exposed cylinder bore surfaces. The produced surfaces consist of randomly scattered pores which provide the surface functionality similar to conventional honing grooves. These pores are formed by uncovered graphite grains. Currently, the automated quality evaluation based on 2D images is still lacked for such surfaces. In this thesis, a morphological feature for describing a set of surface components is proposed. On this basis, a scheme for segmenting graphite grains is designed. This approach enables robust detection of graphite grains even when inhomogeneous illumination and textual disturbances exist.

Moreover, the proposed methods can assist the assessment of surface qualities. The detection results are utilized for constructing quality parameters. Nowadays, several quality parameters [2, 14, 80] for honed surfaces have been implemented for the quality control. Another two parameters are proposed for characterizing the defect severity and the uniformity of graphite distribution, in which engine producers are increasingly interested. Defects and graphite gains can be automatically evaluated in use of the novel algorithms and quality parameters.

## 1.3 Thesis organization

This work is to gain insight into detection-based surface evaluation, which concerns exactly localizing honing defects as well as segmenting uncovered graphite grains. These topics originally stem from quantitative metallography [22, 61], which studies the physical structure and components of metals. Physical features of the material, the mechanical process and the imaging mechanism play essential roles for the surface structures shown in images. In order to understand the inspection tasks correctly, it is necessary to study in depth the surface functions and manufacturing processes. Chapter 2 will explicate these issues in detail.

In Chapter 3 the acquisition methods of micrographs are introduced. Image qualities are tuned to the satisfactory status for displaying relevant surface components in surface finish. Surface structures can be visually recognized based on their shading. The illustrated pictures were captured in motor factories. Some exemplars of inspection objects have been identified by experts, which helped the understanding of inspection tasks. The algorithms proposed in this thesis will detect and evaluate these objects automatically.

Chapter 4 presents a modification of the structure tensor for analyzing local orientations. Furthermore, this method for the orientation analysis is applied to defect detection on plateau-honed surfaces. Surface defects are to be localized in SEM images that are well configured at a high magnification. Since surface components are spatially occluded, it is enabled to regionally segment defects and honing grooves. Technically, feature-based detection scheme is developed to reveal defective locations in SEM images. The weighted orientation dispersion is taken as the signature of defective edges. The algorithm for orientation estimation combines the classic gradient-based method with an edge-preserving filter. The experiments will show that the detection scheme features both an improved robustness and accuracy.

In Chapter 5 a multiscale morphological feature for segmenting uncovered graphite grains is developed for inspecting laser-exposed cylinder liners. LOM images with a low magnification are appropriate for studying the distribution of graphite grains. On machined surfaces, a non-uniform illumination and varying background textures may have a negative impact on the stability of graphite detection. Conventional methods relying on image intensities are unable to yield correct segmentations. The work begins with the study of morphological shapes of the image topography. It can be observed that the interpretation of intensity peaks and valleys is related to observation scales. Inspired by these findings, surface components at different roughness levels can be assigned to fore- and background images. Graphite grains are accurately detected by post-processing of the binarized foreground map.

Chapter 6 proposes quality parameters describing the defect severity and the distribution of graphite grains. Detected surface components are statistically analyzed in this chapter. By testing a series of surface samples, quantitative evaluations are compared with the visual impression.

# 2 Honed surfaces

## 2.1 Cylinder manufacturing

The material and the manufacturing method are two critical factors affecting cylinder qualities. Since cylinder bodies must be able to withstand the internal high-temperature and high-pressure environment, the cylinder material is required to possess enough hardness and toughness. Additionally, uncovered material ingredients are incorporated into surface textures. The commonly used cylinder materials [65] include cast iron with lamellar graphite (GJL), cast iron with vermicular graphite (GJV) and aluminum-silicon alloy. GJL is a traditional material for engine cylinders. Recently, GJV is considered to be an alternative material due to its excellent mechanical and physical properties. However, the production cost of GJV is higher than that of GJL. In contrast to cast iron, aluminum alloy gradually becomes popular, since it can considerably reduce the engine weight. For the sack of the inadequate cylinder strength, the application of the aluminum material is still confined to low power engines. Due to the respective merits and drawbacks of these materials, they co-exist now with each other in the engine world.

Honing [35, 67–69] is an abrasive machining method, typically applied to cylinder finishing. Honing tools are equipped with honing stones that feature an abrasive material, such as silicon carbide or diamond. As illustrated in Fig. 2.1, the honing process consists of rotating and sliding the honing tool. The velocity of honing stones can be decomposed into the tangential and axial components. By adjusting these two speeds, it is convenient to designate the honing angle according to the following formula [21]:

$$\alpha = 2 \tan^{-1} \frac{v_2}{v_1}.\tag{2.1}$$

In this way, honed surfaces are structured with two sets of parallel grooves. Such a texture is technically called "cross hatch". Modern honing processes can achieve highly precise bore dimensions through multiple stages. The grain size of honing stones should be carefully chosen

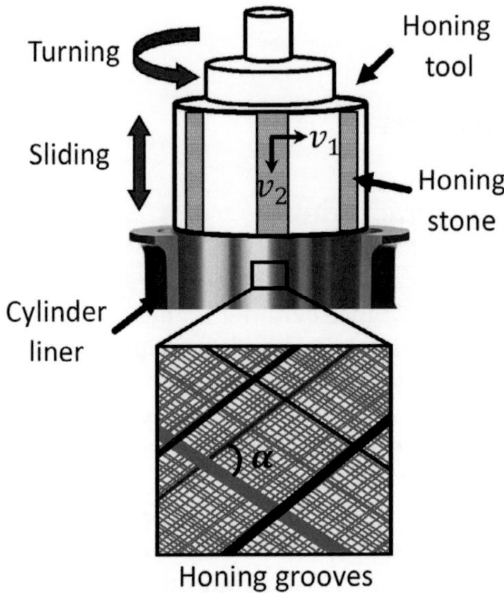


**Figure 2.1** Honing process.

in each honing phase. The smaller crystal grains honing stones have, the finer grooves can be machined, that is, the smoother bore surfaces can be produced. Therefore, the shape and the roughness of cylinder bores have a close relation with honing stones.

For cylinders made of cast iron, honing with coarse abrasives has two effects: honing stones expose graphite components by cutting off the material close to the surface; in the meantime they smear burrs into metal flakes along cutting edges. In consequence, opened graphite particles are enclosed again by the folded metal. To improve surface roughness the metal folds appearing as sharp peaks are required to be removed by fine honing. By sophisticated control of machine parameters coarse tool marks cannot be worn off thoroughly. In this case, grooves of different scales are overlapped in the surface finish. Since deep grooves partition the surface into diamond-shaped flat areas, this method is technically referred to as "plateau-honing" [18]. Alternatively, if merely the finest honing grooves are preserved, the method corresponds to "peak-honing" [18]. The recent advances in honing technology are made by integrating laser treatments,

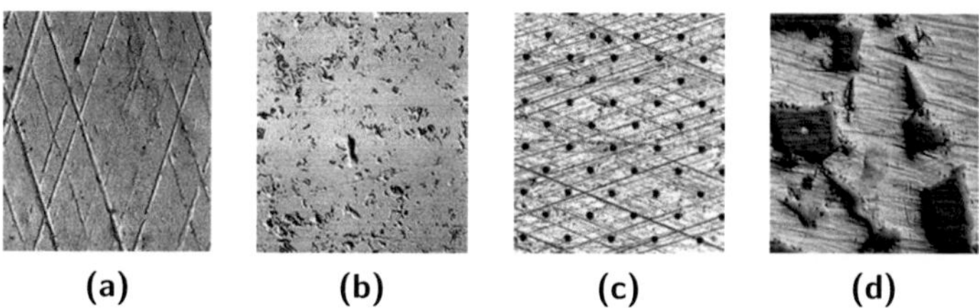

**(a)**    **(b)**    **(c)**    **(d)**

**Figure 2.2**  Honing textures (courtesy of [68]). (a) Plateau-honed. (b) Laser-exposed. (c) Laser-honed. (d) MMC-casted.

for instance, deliberately engraving regular pocket structures on plateau-honed surfaces [35] or by laser-exposure of peak-honed surfaces aiming to melt a layer of metal folds and consequently open graphite grains [68]. For aluminum cylinders, honing is used to expose metal matrix composites (MMC) [13]. Textures obtained in aluminum cylinder liners are completely different from those produced with cast iron. In terms of aforementioned honing methods four types of honing textures are available in practice: plateau-honed, laser-exposed, laser-honed and MMC-casted. See also Fig. 2.2. This thesis will focus on the first two classes of honing textures. The tribological sense behind these textures will be explained in the rest of this chapter.

## 2.2  Surface function

During engine operation, the fuel energy is converted into the kinetic energy of pistons. Due to the friction between the cylinder wall and the piston ring, energy losses are inevitable. An effective means to reduce friction is to make use of a lubrication mechanism to isolate contact surfaces. The formation of the lubricating layer, that is, an oil film, is a hydrodynamic result. Motor oil arrives at bore surfaces via the channel inside the piston. With the movement of the piston, the whole cylinder wall can be lubricated by the oil. The thickness and the distribution of the oil film highly depend on the structure of the cylinder bore surfaces. At the roughness scale, valleys in the surface profile serve as oil-reservoirs, from which the piston ring obtains sufficient lubricant for pulling out a fluid film. In addition to the influence on lubrication, bore surfaces are also tightly related to oil consumption and noxious emission. When the pis-

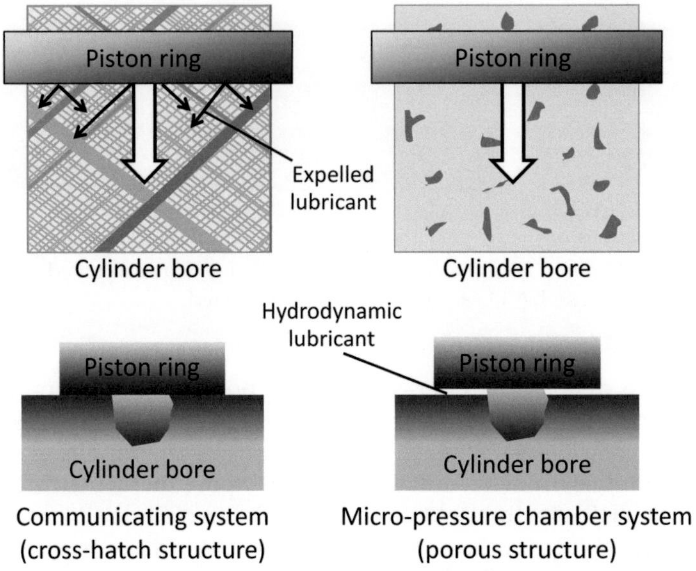


**Figure 2.3**  Tribological systems (courtesy of [35]).

ton ring runs back to the bottom of the cylinder liner, motor oil should be scraped into the crankcase oil tray for the repeated use. Otherwise the remaining oil will be combusted together with the fuel. In this case, $CO_2$ emission will increase dramatically. The experts' experience tells that oil consumption is mainly caused by "oil leakage" [18]. Peaks on rough bore surfaces do not only bring about strong piston wear, but also damage cylinder walls by leaving axial grooved traces. Thereby, lubricant may enter the combustion chamber along large gaps. For these reasons the desired surface structures of cylinder bores should be featured with small roughness on the part contacting the piston while also guarantee the adequate oil reserve in valleys.

Modern honing technologies allow producing functionally designated textures. For example, on plateau-honed surfaces lubricant is dispersed in honing grooves, which constitute a channel system [35] communicating at intersection points (Fig. 2.3 left panel). The design of the honing angle, $\alpha$, depends on the consideration that motor oil should be evenly distributed in the tangential and axial directions. The latest techniques illustrate that a flat honing angle ($\alpha < 30°$) leads to less oil distribution in the axial direction, and thus the piston tends to be hard-going on the running surface. With an increased honing angle ($\alpha > 90°$), lubricant dominates in the axial direction. In this case too much oil goes into the combustion chamber. To obtain the optimum lubricating effect the honing angle should be chosen in the interval between $30°$ and $90°$. The honing technique could be combined with laser exposure. The right panel of Fig. 2.3 shows that the micro-pressure chamber system [35] substitutes the channel system to retain motor oil. Vermicular graphite is opened by melting metal folds for the purpose of producing porous structures. As motor oil resides in pores, the piston seems to be "swimming" on the surface. Such a structure overcomes the major drawback of the communicating channel system, that is, the friction is increased again after the piston expels oil to the surrounding. At this point it becomes clear why the distribution of opened graphite particles is crucial for surface lubrication. Keeping these surface functions in mind, novel solutions for the inspection of some challenging surfaces are explored in the following parts of this thesis.

# 3 Image acquisition

In the industrial environment, image acquisition of honed surfaces follows internationally standardized metallographic approaches [61, 75]. Light optical microscope (LOM) and scanning electron microscope (SEM) are the major tools used for qualitative and quantitative analysis of honed surfaces. In this thesis surface data are collected from specimens that are prepared by splitting a cylinder liner. The cylinder wall is cut into stripes or small pieces. To obtain a comprehensive understanding of multiscale surface structures, examinations using LOM and SEM images complement each other in the aspects of overall surface appearances and material details.

## 3.1 Light optical microscopy

The LOM [11, 63] is the most efficient and economic instrument compared to any other examination technique in metallography. It takes advantage of a compound lens to focus light into the eyes or a camera. On metallic surfaces the interaction between the specimen and the incident light is recognizable, when light reflected from a surface region (the field of view, FOV) hits the optical system. At the end of the light path an CCD or CMOS sensor converts the light signals into digital images. In Fig. 3.1, the structure of LOM systems is schematically illustrated. Figure 3.2 shows a LOM acquisition for a piece of laser-exposed cylinder bore surface.

Current optical microscopes possess sufficient resolving power for imaging graphite grains. Based on optical geometry, the optical magnification of LOMs is the product of the powers of the ocular and objective lenses. Nowadays, the maximum magnification of LOMs can reach values higher than 1000. Moreover, as the light from a point on the specimen passes through the back aperture of objectives, a disk-like fringe pattern appears in the image plane. Optical resolution is the smallest distance between two points on a specimen that can be distinguished as separate entities. Currently, the maxmal spatial resolution of LOMs is equal to 0.2 $\mu$m. These physical properties of LOM systems can fully meet the requirement for the detection of graphite grains. Normally, uncovered graphite grains

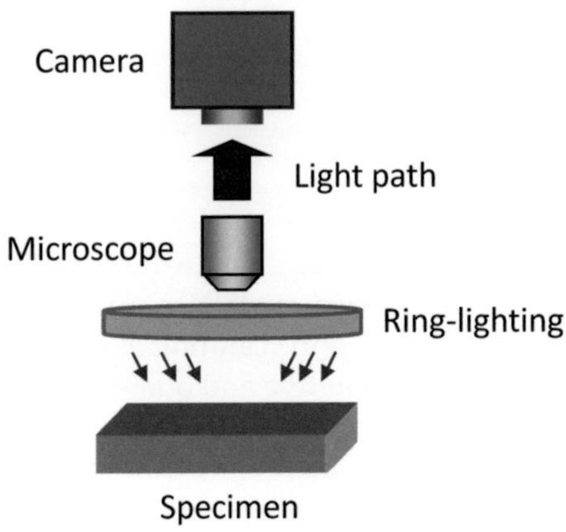



**Figure 3.1**  Light optical microscope system.

have a mean area of ca. $0.2 \times 0.2\,\mathrm{mm}^2$. In Fig. 3.2 every square milimeter contains approximately 10 grains. Total 200~300 grains can be observed in the FOV. Such a magnification is very suitable for the study of grain distribution. Additionally, Fig. 3.2 shows an image block captured by a high-resolution CCD camera. The size of each graphite grain is equal to an image area of 50~100 pixels. This makes graphite grains distinct to other surface components with different sizes. Another property of LOMs that should be mentioned is the depth of focus (DOF). Only those surface points distancing from the focused plane within a tolerence can be sharply imaged by a LOM. Specimens of cylinder bores have curved surface profiles. The DOF should be improved with a low optical magnification or a small numerical aperture [6] in order to obtain a good image quality. By sophiscatedly configuring the LOM, most part of the cylinder bore surface shown in Fig. 3.2 is sharply imaged. The distoration is not recognizable, since the imaged surface patch can be approximated as a plane, when the radius of the cylinder bore (100 mm) is far greater than the extension of the surface patch. Besides, image contrast may be inhomogeneous due to the non-uniform illumination. Such disturbance degrades the image quality. In this inspection task, the qualified surface finish should show fine plateau grooves and evenly uncovered graphite particles. The dark graphite grains shown in this image are to be detected with the method described in Chapter 5.

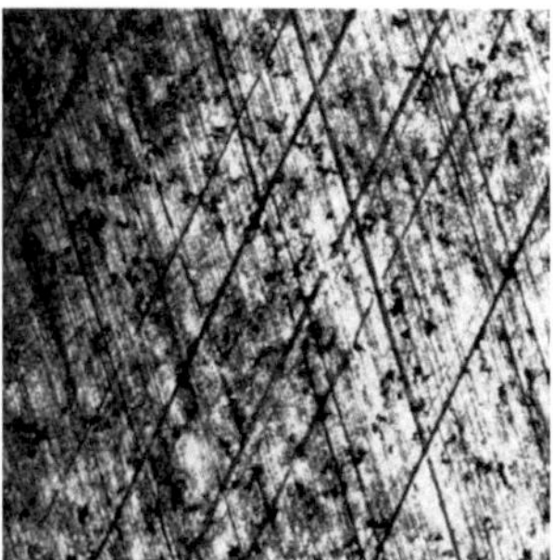

**Figure 3.2** Unqualified laser-exposed cylinder bore surface (evenly uncovered graphite grains, too many deep grooves). FOV $\approx$ 5 × 5 mm$^2$. The image size is 256 × 256.

## 3.2 Scanning electron microscopy

In metallography, material details are preferably examined with a scanning electron microscope. A SEM imaging system [62] is schematically illustrated in Fig. 3.3. It utilizes a high-energy electron beam, namely the primary electrons (PEs), for scanning the specimen. By means of magnetic and objective lenses, the electron beam emitted from the electron gun is focused to a spot on the specimen surface. By altering the electrical current in deflection coils, the electron beam can be deflected with a certain angle. In this way, the scanning spot translates to the next scanning point in a predefined raster pattern. Whenever the electron beam hits the surface, the energy exchange between incident electrons and specimen atoms creates a variety of signals, among which secondary electrons (SEs) are most frequently used for imaging. SEs are released from the surface by excitation. A special detector collects SEs scattered in the sample chamber, and then converts them into successive electrical signals. Furthermore, a signal processing system connected to the detector serves for signal amplification and digitalization. The output signal can be assigned to an image pixel in form of a gray value. When the scanning raster is associated with the image grid, the surface region within the scanning raster is ultimately able to be viewed as a digital image.

Unlike the LOM, the magnification of a SEM is not derived from the power of lenses but from the following relation:

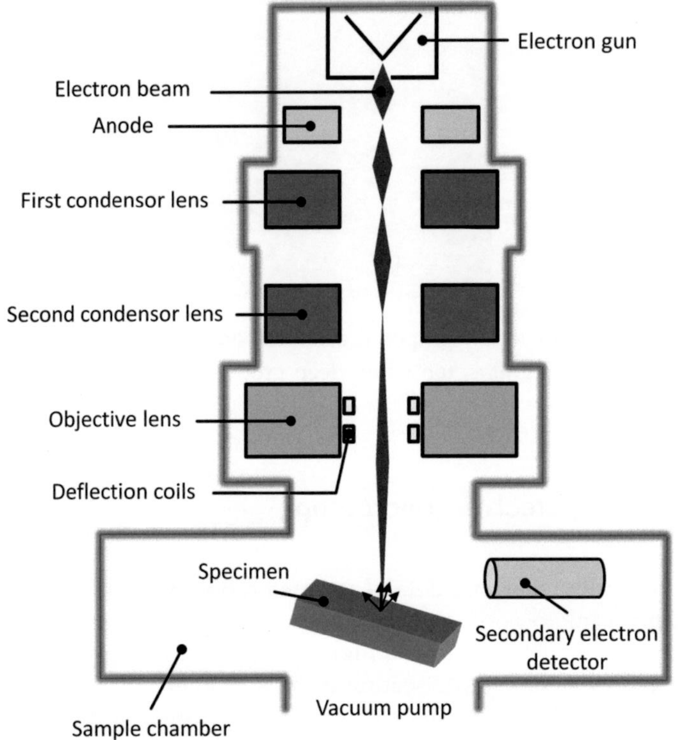


**Figure 3.3**   Scanning electron microscopy.

$$\text{SEM magnification} = \frac{\text{displayed image size}}{\text{raster size on the specimen}}.$$

The lenses of the SEM are only to focus the electron beam into a spot. Assuming that the image size is fixed for the display, the specimen will be enlarged as the raster size is reduced. Accordingly, the FOV becomes smaller. As long as the spot is small enough to resolve the scanning raster, the specimen can be sharply imaged. The brightness of SEM images relies on the amount of collected secondary electrons. Regardless of the primary electron beam and the specimen material, the surface pose is the main factor affecting the emission of secondary electrons. Figure 3.4 demonstrates that the "escape" path turns shorter as the incident angle $\phi$ increases. It

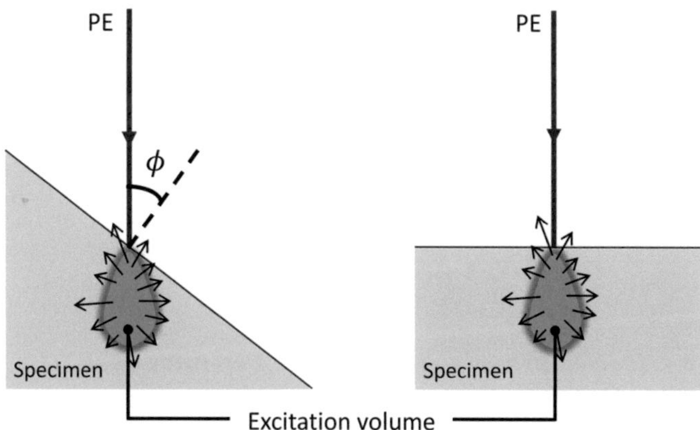


**Figure 3.4** Secondary electron emission. As the primary electron beam enters the specimen, secondary electrons escape from a teardrop-shaped excitation volume.

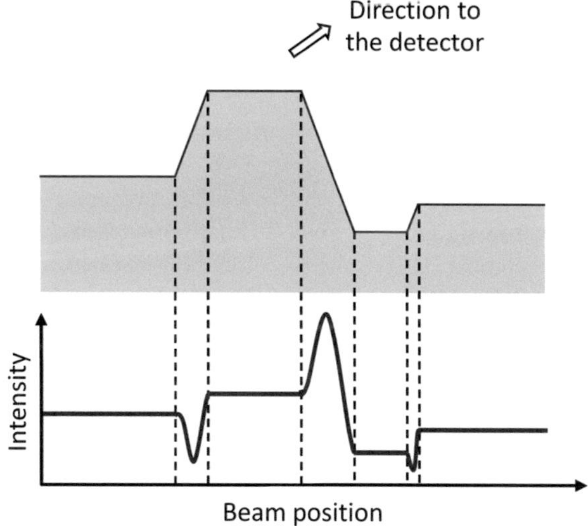


**Figure 3.5** Topography contrast.

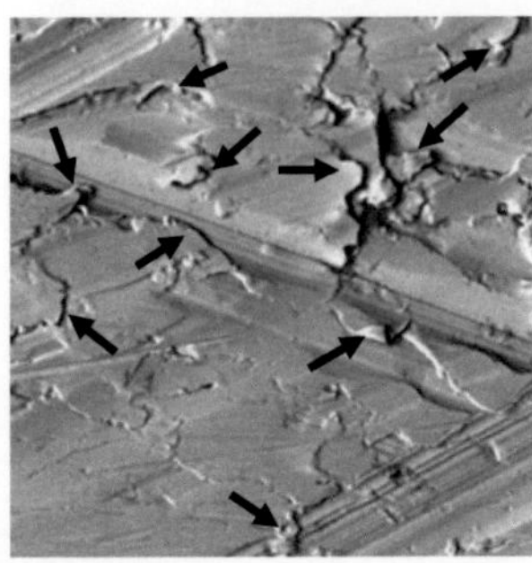

**Figure 3.6** Unqualified plateau-honed surface (too many metal folds, less evident plateaus). FOV $\approx 0.12 \times 0.12 \, \text{mm}^2$. The image size is $256 \times 256$.

means that secondary electrons are easier to emit from steep surfaces. Another phenomenon is that more electrons can be captured from a surface toward the detector than that opposite to the detector. Considering a rough surface profile composed of a number of small sections, the intensities could vary like the schematic diagram shown in Fig. 3.5.

Figure 3.6 displays a piece of a unqualified plateau-honed surface. Owing to the high resolution and the large DOF, the SEM image shows a stereo effect that is not available in LOM images. These properties are quite beneficial for the structure-oriented analysis. This SEM image is later employed for the development of image analysis algorithms. This surface sample shows that the cross-hatch patterns are not successfully machined. Honing grooves are seriously interrupted into fragments. Moreover, it can be observed that the defective locations marked with arrows are folded metal flakes. Such a surface structure will induce a strong surface friction. The piston ring is also likely to be damaged by sharp metal folds. In addition, plateau grooves do not abound in this image. As a result, the piston is hard-going on such a surface. The algorithm for defect detection will be presented and experimentally validated with this image in Chapter 4.

# 4 Defect inspection

For plateau-honed cylinders, the presence of two bands of honing grooves is desired in the manufacturing process. Due to the imperfectness of metalworking, honing grooves are smeared and interrupted by folded metal. These material defects result in frictional losses and accelerate the wear of the piston. Moreover, exfoliated metal flakes increase the particle emission in the running process of engine operation. Currently, the grading of defect severity is still a demanding work. In most cases, metal folds are visually analyzed in SEM images. The obtained inspection results are not reliable. For these reasons, quantitative evaluation of surface qualities is in demand.

Plateau-honed surfaces possess a physically inhomogeneous topography. Two roughness levels exist in s honing textures. The surface components with low roughness are considered as smooth plateaus, on which fine grooves show a very low image contrast. These smooth surface regions are partitioned by deep honing grooves, which feature a high roughness level. SEM images can faithfully reflect the intrinsic surface topography in a 2D fashion. Deep grooves are well contrasted in gray-level SEM images. Surface damages like metal folds exist on both roughness levels. These manufacturing failures are distinct, when they extend beyond the surface. Weak scratches on plateaus have little influence on engine properties. Hence, the presented detection will concentrate on the high roughness level, which shows higher image contrast than that at plateaus. In this chapter the state-of-the-art approaches for detecting metal folds are reviewed firstly. Afterwards, a new inspection strategy based on the edge-aware structure tensor is proposed. Compared with previous works, the method is edge-based and independent of honing grooves. It will be shown that the presented method is applicable to inspect honed surfaces owning a wide range of qualities.

## 4.1 Previous work

### 4.1.1 Approach using signal processing

The early work for the detection of metal folds ("Blechmantel" in German) and other defects (holes, groove interrupts, stains, etc.) was presented by Beyerer and Puente [3, 58]. Their image analysis approach is based on a signal model. Ideal honing texture is assumed to consist of compounded groove sets according to the designated spatial geometry. Surface defects are treated as abnormal places in groove textures. In the last few years this model has been the foundation of analyzing honed surfaces in digital images. The classic application of Beyerer's model was separating straight lines from an isotopic background. The framework was originally developed in the Fourier domain [5]. Recently, an improvement was made in [80], where the wavelet transform was utilized to optimize the textural decomposition in WLI images.

The family of algorithms following Beyerer's model shows a consensus that defect inspection should become trivial in the separated background image. Here Beyerer's scheme is briefly recalled in Fig. 4.2. Figure 4.1 shows the detection result obtained by using the test image shown in Fig. 3.6. The 2D Fourier coefficients representing textures consisting of straight grooves are distributed along compact radial lines across the origin. The background is assumed to be isotropic and, according to the Riemann-Lebesgue lemma, its spectrum concentrates around the origin of the spatial frequency domain. After splitting the image spectrum into an anisotropic and an isotropic part, the groove texture and the back-

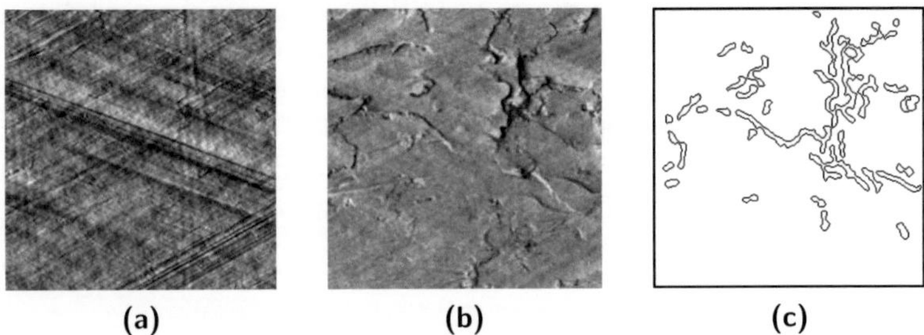

(a)          (b)          (c)

**Figure 4.1** Signal decomposition of Fig. 3.6. (a) Groove texture. (b) Background. (c) Defective edges.

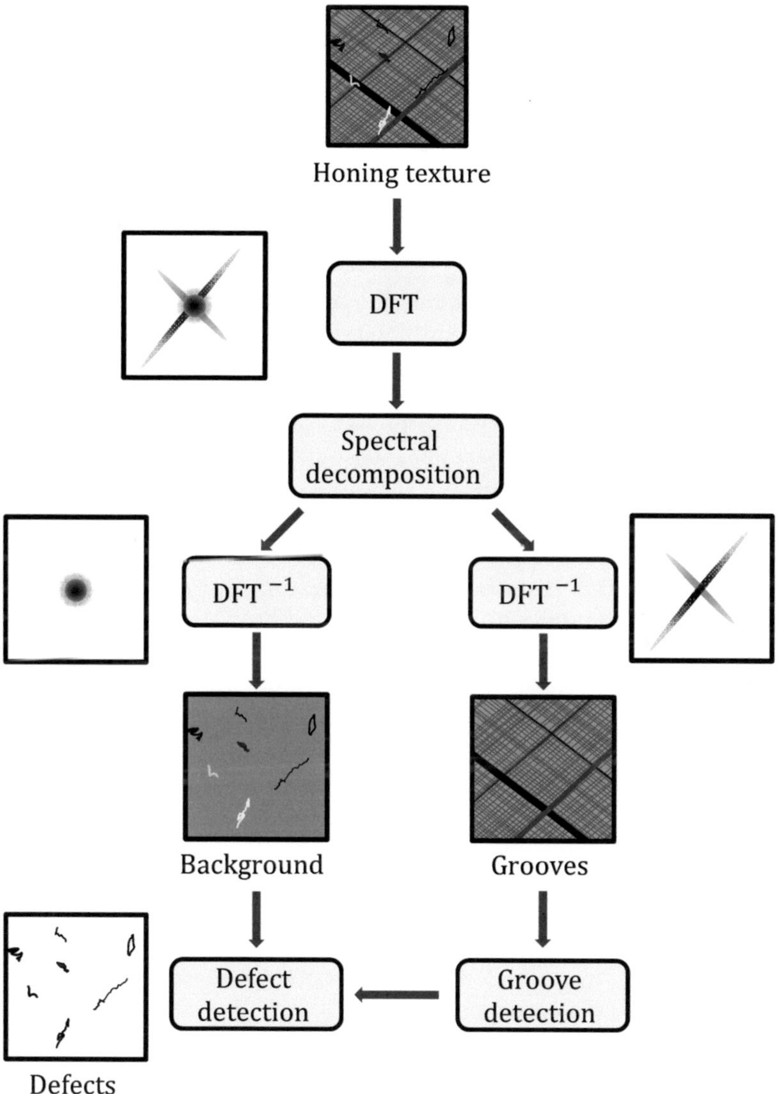



**Figure 4.2**   Beyerer's scheme for defect detection.

ground can be reconstructed by using the inverse Fourier transform. Defects can be localized in the background image by edge detection. To obtain closed contours of defective edges, Laplacian-of-Gaussian (LoG) filters are adopted in their work. Their results can be further optimized by limiting the detection within grooved regions.

The model-based textural decomposition turns out to be problematic for analyzing seriously defective surfaces. The spectral analysis of honing textures handles surface images globally. The algorithm works well for honed surfaces in which grooves are present throughout the entire image. Practical manufacturing processes may be disturbed by many complex factors. Machine parameters, the material of honing stones and the coolant could lead to significantly varying surface qualities. The test image shown in Fig. 4.1 illustrates that the global algorithm is not capable of thoroughly eliminating groove fragments from the background. Artifacts may appear in separated images, because grooves reconstructed from the 2D Fourier domain are intact lines without gaps. The presence of groove interrupts limits the anisotropic character of grooves, which may induce an improper assignment of Fourier coefficients during the spectrum decomposition. Moreover, it should be argued that the influence of image noise is not ignorable in the stage of detection. During the LoG-filtering, unexpected zero-crossings could be created by noise. As a result, defect contours may be linked with noise contours. Such a problem is troublesome, since defects could be incorrectly localized. So far, the exact finding of defects is still challenging for 2D image analysis of honed surfaces. In view of the drawbacks of global signal decomposition, the algorithm developed in this thesis should be independent of the presence of honing grooves.

### 4.1.2 Approach based on surface measurement

The height of surface profiles provides additional information for surface inspection. In general, the surface topography can be separated into material spurs, plateaus and valleys [78, 80]. The assessment for material defects not only depends on surface heights but also on textural features. Xin [80] and Dimkovsiki [14] brought forward similar interpretations of honing defects that are observed with a white light interferometer (WLI). They believed that tool marks on honed surfaces should be non-interrupted grooves so that motor oil can flow smoothly. This family of methods segmented a 2.5D image into deep grooves and plateaus.

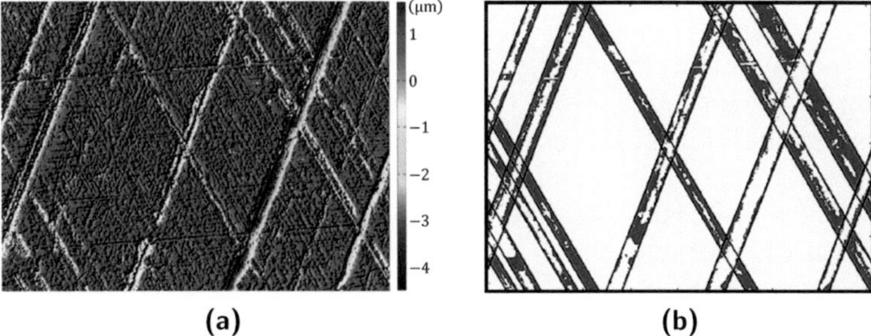

**Figure 4.3**  Detection results published in [14]. (a) WLI image of a plateau-honed surface. (b) Coarse grooves interrupted by metal folds that are colored in gray.

Boundaries of coarse grooves were determined by the Radon transform. Plateaus were regions among coarse grooves. Based on surface heights, they found blocked regions inside globally reconstructed grooves. These groove interruptions were detected as metal folds. Furthermore, peaks rising on plateaus were regarded as material burrs. This idea is illustrated in Fig. 4.3. However, the approaches mentioned above suffer from some drawbacks:

- The defect detection depends on the estimation of groove parameters, such as the position, the width and the angle of individual grooves. Unfortunately, it is difficult to estimate these parameters at smeared grooves. The long-distance interaction of image contents is often needed to link groove fragments.

- Nowadays, some motor producers do not consider interrupted grooves as manufacturing failures, but treat them as a part of the micro-press chamber system. Such an understanding conflicts with the definition of defects advocated by Xin and Dimkovsiki.

To avoid confusion, only the defects that extend beyond the surface are taken into account. Metal folds that have been pressed into surfaces are not in the scope of the study, because they do not really increase surface friction. The approach identifies honing failures at defective edges which have irregular shapes due to the smearing of honing stones.

## 4.2 Local orientation analysis

In 2D gray-level images, strong intensity variations are referred to as image edges. The locations where the intensity variation happens form spatial edge shapes. Physical groove boundaries are imaged as straight edges of high strength, while metal folds are distinct due to their sharp and rough edges. In nature, straight and rough edges can be distinguished with local orientations. Thus, local orientation analysis is the main techinique used in the presented detection scheme.

In the literature the orientation information has been known as a significant feature of oriented textures. A varity of orientation operators was presented. The choice of orientation operators is quite dependent of applications. Some special properties of textures were emphasized in the process of feature extraction. In this context these methods are sorted into two classes, that is, gradient-based and filter-based methods. In the following, the characteristics of these methods are reviewed. Then, a novel approach is introduced in Section 4.3.

### 4.2.1 Filter-bank based methods

#### 4.2.1.1 A short survey

Filter-bank based algorithms look for directional structures with scaled and rotated filter kernels. The selected filter bank can be composed of Gaussian derivative filters [37], Gabor filters [43] or steerable filters [20]. Filtered images construct an orientation space [9, 23] or orientation-scale space [84], in which overlapped objects with multiple orientations can be segmented. Since each kernel of the filter bank only covers limited spatial frequencies, high-frequency noise could be discarded from filter channels. Filter-bank based methods are thus more resistant to noise than gradient-based ones. The prominent weakness of these methods is the limited orientation accuracy due to the fixed number of filters.

#### 4.2.1.2 Gabor-filter bank

As reported in [1], the Gabor-filter bank is an oriented feature detector with good performance. Through creating an orientation-scale space, many useful edge features can be extracted. Inspired by this, in Section 4.3.3 a set of feature-adaptive filter kernels is developed to improve the estimation of local orientations. Several techniques [25, 38, 73] have been reported in the literature for the design of Gabor filters. The advice of Manjunath and Ma [38] is adopted here, becase their method has been validated in many applications on texture segmentation and object recognition. A 2D complex Gabor function and its Fourier transform can be formulated as

$$f(x,y) = \frac{1}{2\pi\sigma_x\sigma_y}\exp\left[-\frac{1}{2}\left(\frac{x^2}{\sigma_x^2} + \frac{y^2}{\sigma_y^2}\right) + 2\pi j u_0 x\right], \qquad (4.1)$$

$$F(u,v) = \exp\left\{-\frac{1}{2}\left[\frac{(u-u_0)^2}{\sigma_u^2} + \frac{v^2}{\sigma_v^2}\right]\right\}, \qquad (4.2)$$

where $\sigma_u = 1/2\pi\sigma_x$ and $\sigma_v = 1/2\pi\sigma_y$. The band limit of a Gabor filter is defined with its half-peak magnitude support in the Fourier domain. The Gabor-filter bank is a class of self-similar Gabor functions that are generated by scaling and rotating the basic filter $f(x,y)$. The members of Gabor-filter bank are non-orthogonal. In order to reduce the redundancy in filtered images, the filter supports in the Fourier domain are aligned to touch each other. Given the number of orientations, $N_\xi$, and the number of scales, $N_\eta$, a Gabor filter bank can be notated as $\left\{f_{\xi,\eta}(\mathbf{x})\right\}$ with $\xi = 1, 2, \ldots, N_\xi$ and $\eta = 1, 2, \ldots, N_\eta$. An example is shown in Fig. 4.4.

Formally, the member of a Gabor-filter bank can be written as

$$f_{\xi,\eta}(x,y) = \rho^{-\eta} F\left(x',y'\right), \qquad (4.3)$$

with

$$x' = \rho^{-(\eta-1)}\left(x\cos\theta + y\sin\theta\right),$$
$$y' = \rho^{-(\eta-1)}\left(-x\sin\theta + y\cos\theta\right),$$

The parameters of Equ. 4.3 are calculated with following formulas:

$$\theta = \frac{(\xi - 1)\,\pi}{N_\xi},$$

$$\rho = \left(\frac{u_{N_\eta}}{u_1}\right)^{-\frac{1}{N_\eta - 1}},$$

$$u_0 = u_{N_\eta},$$

$$\sigma_u = \frac{(\rho - 1)\,u_{N_\eta}}{(\rho + 1)\,\sqrt{2\ln 2}},$$

$$\sigma_v = \tan\left(\frac{\pi}{2N_\xi}\right)\left[u_{N_\eta} - 2\ln\left(\frac{\sigma_u^2}{u_{N_\eta}}\right)\right]\left[2\ln 2 - \frac{(2\ln 2)^2\,\sigma_u^2}{u_{N_\eta}^2}\right]^{-\frac{1}{2}}.$$

The filtered images are complex, and can be written as

$$\breve{g}_{\xi,\eta}(\mathbf{x}) = g_{\xi,\eta}(\mathbf{x}) * f_{\xi,\eta}(\mathbf{x}). \tag{4.4}$$

It is worth to note that in the spatial domain the real (odd) part of complex Gabor filters is capable of detecting step edges, while the imaginary (even) part is useful for finding roof edges. Since both step and roof edges should be detectable for the inspection of honed surfaces, the Gabor energy function is utilized for analyzing the structural anisotropy. Hence, a matrix of size $N_\xi \times N_\eta$ is constructed at a pixel location, $\mathbf{x}$. The matrix element is $\left|\breve{g}_{\xi,\eta}(\mathbf{x})\right|^2$. As explained in [34], $\left|\breve{g}_{\xi,\eta}(\mathbf{x})\right|^2$ will be high when $f_{\xi,\eta}(\mathbf{x})$ coincides with the edge width and the edge angle. To gain scale-invariant edge features, the maximum of $\left|\breve{g}_{\xi,\eta}(\mathbf{x})\right|^2$ is chosen from the dimension that repesents the scales [33]. The resulting feature is merely related to orientations:

$$\tilde{g}_\xi(\mathbf{x}) = \max_\eta \left|\breve{g}_{\xi,\eta}(\mathbf{x})\right|^2. \tag{4.5}$$

The values of $\tilde{g}_\xi(\mathbf{x})$ at all orientations indicate an energy distribution, which is able to describe local multi-oriented structures.

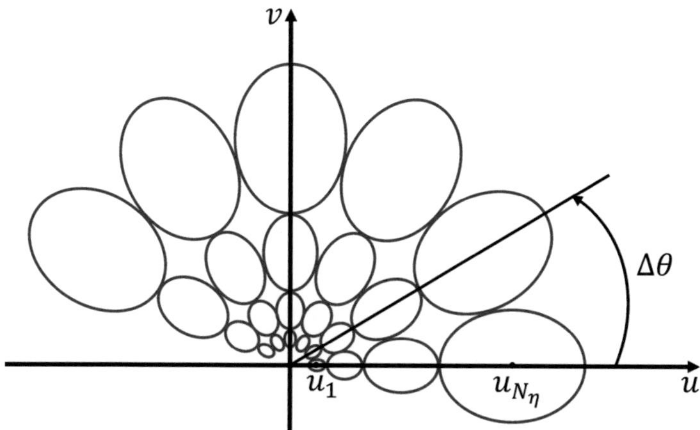


**Figure 4.4**  Gabor-filter bank in 2D Fourier domain. $N_\zeta = 6$ and $N_\eta = 4$. $u_1$ and $u_{N_\eta}$ are the lowest and highest center frequencies, respectively. $\Delta\theta$ is the angle interval.

## 4.2.2 Gradient-based methods

### 4.2.2.1 A short survey

Gradient-based approaches utilize the structure tensor [29] to estimate local dominant orientations. The resulting orientations deliver higher accuracy at fine scales than filter-bank based methods. However, conventional derivative filters used to compute gradient vectors are sensitive to image noise. For this reason, the estimation accuracy of local orientations is often degraded due to image noise. To deal with this problem, some researchers improved the robustness to noise by using alternative computational methods. The orientation operator proposed in [32] implements a band-pass filter by integration. Another author [40] approximated the structure tensor by auto-correlation. Some efforts were also made to improve the structure tensor by adaptively varying the smoothing mechanism [7, 8, 15, 46, 81]. Besides, many applications [31, 41, 55] require to smooth the orientation field without destroying relevant singular points. Diffusion-based regularization [10], model-based optimization [83] and block-wise voting schemes [77] are approaches that were specially designed for this purpose. The resulting orientation fields were coarsened in order to describe objects at desired scales.

In this thesis, the detection of metal folds is associated with edge orientations. The idea is to distinguish rough and straight edges via a descriptor of edge shapes, which requires high accuracy not only for edge angles, but also for edge locations. The conventional gradient-based approach is considered to be suitable for this applcation, and thus chosen as the foundation of this work. Firstly, this approach is recalled. Its problems are illustrated through some tests. To improve the robustness, the gradient-based approach is then optimized in Section 4.3.

### 4.2.2.2 Classic structure tensor

The optimal orientation estimation is assumed to fulfill the least squares principle. The question can be formulated as minimizing the following objective function in terms of the unity vector $\mathbf{n}$ representing the optimal orientation:

$$J = \frac{1}{N^2} \sum_{\mathbf{x} \in \Omega_{\mathbf{W}}} \left| (\nabla g(\mathbf{x}))^{\mathrm{T}} \mathbf{n} \right|^2. \tag{4.6}$$

$\mathbf{x} = (x, y)^{\mathrm{T}}$ denotes a pixel location. $\nabla g(\mathbf{x}) = \left( g_x, g_y \right)^{\mathrm{T}} (\mathbf{x})$ is the gray-level gradient of the image $g(\mathbf{x})$. $\Omega_{\mathbf{W}}$ is the support of a local window, $\mathbf{W}$, which is sized to $N \times N$. The minimization problem can be solved by eigendecomposition of the structure tensor,

$$\mathbf{T_C}(\mathbf{x}) = \begin{pmatrix} T_{11} & T_{12} \\ T_{21} & T_{11} \end{pmatrix} (\mathbf{x}) \tag{4.7}$$

with

$$T_{11}(\mathbf{x}) = \left( g_x^2 * G_{\mathbf{W}} \right) (\mathbf{x}),$$

$$T_{22}(\mathbf{x}) = \left( g_y^2 * G_{\mathbf{W}} \right) (\mathbf{x}),$$

$$T_{12}(\mathbf{x}) = T_{21}(\mathbf{x}) = \left( g_x g_y * G_{\mathbf{W}} \right) (\mathbf{x}),$$

where the structure tensor $\mathbf{T_C}(\mathbf{x})$ is defined as the smoothed dyadic product of the gradient vector, $G_{\mathbf{W}}(\mathbf{x})$ denotes a Gaussian or averaging filter kernel, and $*$ is the convolution operator. Computationally, the eigenvalues can be calculated directly with

$$\lambda_1\left(\mathbf{x}\right) = \frac{1}{2}\left[\det\left(\mathbf{x}\right) + \sqrt{\det^2\left(\mathbf{x}\right) - 4\mathrm{tr}\left(\mathbf{x}\right)}\right], \tag{4.8}$$

$$\lambda_2\left(\mathbf{x}\right) = \frac{1}{2}\left[\det\left(\mathbf{x}\right) - \sqrt{\det^2\left(\mathbf{x}\right) - 4\mathrm{tr}\left(\mathbf{x}\right)}\right]. \tag{4.9}$$

Here, $\det\left(\mathbf{x}\right)$ is the determinant of $\mathbf{T_C}\left(\mathbf{x}\right)$, and $\mathrm{tr}\left(\mathbf{x}\right)$ is the trace of $\mathbf{T_C}\left(\mathbf{x}\right)$:

$$\det\left(\mathbf{x}\right) = T_{11}\left(\mathbf{x}\right) T_{22}\left(\mathbf{x}\right) - T_{12}^2\left(\mathbf{x}\right).$$
$$\mathrm{tr}\left(\mathbf{x}\right) = T_{11}\left(\mathbf{x}\right) + T_{22}\left(\mathbf{x}\right),$$

Moreover, the angle of the eigenvector corresponding to the largest eigenvalue is expressed as

$$O\left(\mathbf{x}\right) = \tan^{-1}\frac{2T_{12}\left(\mathbf{x}\right)}{\left(T_{22} - T_{11}\right)\left(\mathbf{x}\right)}. \tag{4.10}$$

The local dominant orientation is perpendicular to $O\left(\mathbf{x}\right)/2$.


### 4.2.2.3 Property

The mathematical expressions in Section 4.2.2.2 are usually used for the calculation. By smoothing the whole image domain, the mean squared-gradient field can be obtain as follows:

$$\mathbf{u}\left(\mathbf{x}\right) = \lambda_1\left(\mathbf{x}\right) e^{jO\left(\mathbf{x}\right)}.$$

The pros and cons of the classic structure tensor are illustrated with the test image shown in Fig. 4.5. The window size critically affects the orientation estimation. The large window significantly reduces noise but introduces a strong blurring in both the amplitude and orientation field. In contrary, the rough edge is preserved by using a small window. However, noise reduction may be insufficient at straight edges. Besides, the classic structure tensor can only deal with the intrinsic 1D structure [17] that shows only one dominant orientation. At the structure mixture (e.g., junctions) the estimated orientation will be far from each oriented structure contained in the local window.

To detail the reasons of aforementioned issues, the eigenvalues and eigenvectors of the classic structure tensor are formulated in an alternative form. According to [31], the eigenvalue analysis of the structure tensor is equivalent to averaging squared-gradients in a local window. Let me represent the gradient vector with a complex variable:

$$\nabla g\left(\mathbf{x}\right) = g_x\left(\mathbf{x}\right) + \mathrm{j}g_y\left(\mathbf{x}\right). \tag{4.11}$$

The squared-gradient can be expressed as

$$r\left(\mathbf{x}\right) e^{\mathrm{j}\varphi(\mathbf{x})} = g_x^2\left(\mathbf{x}\right) - g_y^2\left(\mathbf{x}\right) + 2\mathrm{j}g_x\left(\mathbf{x}\right) g_y\left(\mathbf{x}\right). \tag{4.12}$$

Then, squared-gradients in a local window can be visualized on the complex plane with a compass plot. The obtained vector map demonstrates a gradient distribution which can reflect edge shapes. Artificial patterns shown in Fig. 4.6 are used to illustrate four typical structures — straight edges, rough edges, junctions and smooth patches. Gaussian noise is deliberately added to these pictures. The corresponding gradient distributions are depicted in Fig. 4.7. It can be seen that the dispersion of squared-gradients changes with the edge strength and the noise level. Therefore, the mean squared-gradients are also contrast-variant and noise-sensitive.

The performance of the structure tensor can be further discussed in real images. A surface image is represented with the following model:

$$g\left(\mathbf{x}\right) = i\left(\mathbf{x}\right) t\left(\mathbf{x}\right) + n\left(\mathbf{x}\right), \tag{4.13}$$

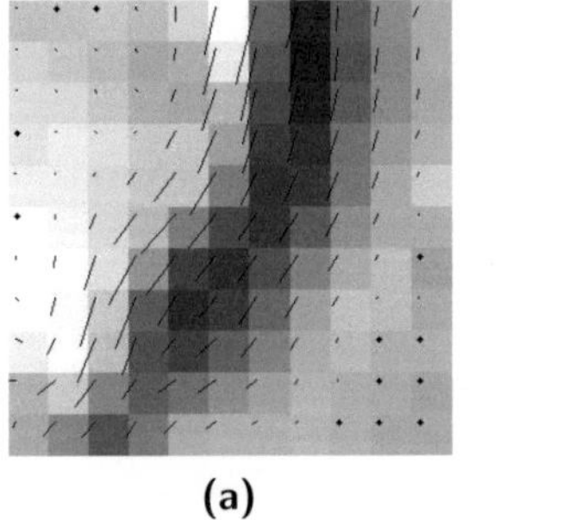 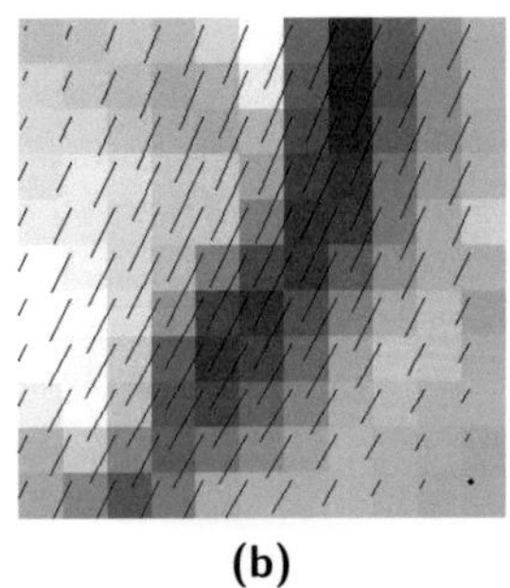

(a)        (b)

**Figure 4.5** Vector fields, $\lambda_1\left(\mathbf{x}\right) e^{\mathrm{j}(O(\mathbf{x})/2+\pi/2)}$, estimated by the classic structure tensor. (a) $N = 5$, (b) $N = 21$.

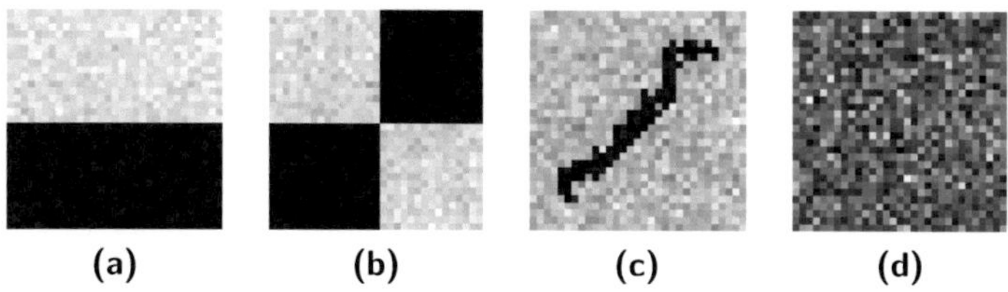

(a)  (b)  (c)  (d)

**Figure 4.6**  Noisy artificial patterns. (a) Straight edge. (b) Junction. (c) Rough edge. (d) Smooth patch.

where $t(\mathbf{x})$ is the undisturbed image signal of a relief texture, $i(\mathbf{x})$ denotes the inhomogeneity, and $n(\mathbf{x})$ is additive white Gaussian noise (AWGN). The gradient field derived from Equ. 4.13 is

$$\nabla g(\mathbf{x}) \approx \nabla i(\mathbf{x}) \nabla t(\mathbf{x}) + \nabla n(\mathbf{x}), \qquad (4.14)$$

where the derivative of $i(\mathbf{x})$ is neglected, because $i(\mathbf{x})$ normally varies very slowly in local regions. Equ. 4.14 means that $i(\mathbf{x})$ scales the gradient amplitude, $|\nabla t(\mathbf{x})|$, but does not tune the gradient angle of $t(\mathbf{x})$. In comparison, image noise distorts both the gradient amplitude and the angle. Let me notate the squared-gradient of $t(\mathbf{x})$ in the polar coordinate system as $r_t(\mathbf{x})\, e^{j\varphi_t(\mathbf{x})}$. Then, the squared-gradient of $g(\mathbf{x})$ can be described with

$$r(\mathbf{x})\, e^{j\varphi(\mathbf{x})} = \left( i^2(\mathbf{x})\, r_t(\mathbf{x}) + r_n(\mathbf{x}) \right) e^{j(\varphi_t(\mathbf{x}) + \varphi_n(\mathbf{x}))}, \qquad (4.15)$$

where $r_n(\mathbf{x})$ and $\varphi_n(\mathbf{x})$ are the biases in the radial amplitude and the rotated angle, respectively. This representation inherits the properties of Equ. 4.14 and combines the signal and the noise in a single term. Furthermore, it should be noted that strong gradients have higher votes to the vector addition than weak ones. This enables to describe local dominant structures by simplifying Equ. 4.15 as follows:

- Squared-gradients at strong edges can be approximately characterized with $i^2(\mathbf{x})\, r_t(\mathbf{x})\, e^{j(\varphi_t(\mathbf{x}) + \varphi_n(\mathbf{x}))}$;

- Weak edges and smooth regions can be expressed as $r_n(\mathbf{x})\, e^{j\varphi_n(\mathbf{x})}$, since these places are dominated by noise.

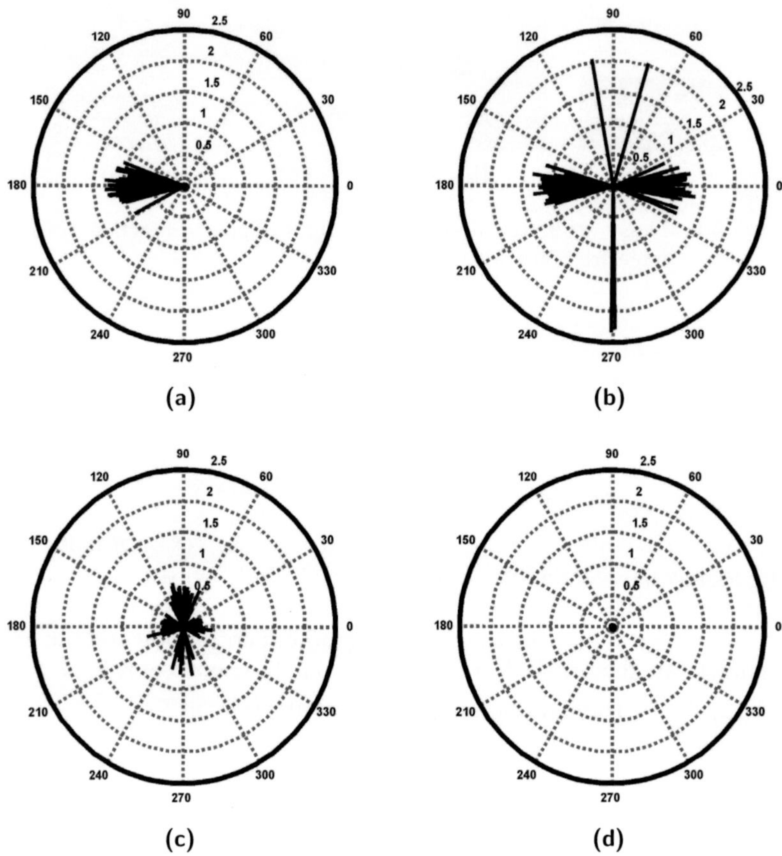

**Figure 4.7**  Gradient distributions of noisy patterns. (a) Straight edge. (b) Junction. (c) Rough edge. (d) Smooth patch.

Based on the analysis above, the mean squared-gradient in a local window can be formulated as

$$R_{\mathbf{W}}e^{j\Phi_{\mathbf{W}}} \approx \frac{1}{N^2} \sum_{\mathbf{x}\in\Omega_{\mathbf{W}}} \left( i^2\left(\mathbf{x}\right) r_t\left(\mathbf{x}\right) + r_n\left(\mathbf{x}\right) \right) e^{j\left(\varphi_t\left(\mathbf{x}\right)+\varphi_n\left(\mathbf{x}\right)\right)}. \qquad (4.16)$$

The decomposition is then accomplished in the following equations:

$$R_{\mathbf{W}}e^{j\Phi_{\mathbf{W}}} = R_{t}e^{j\Phi_{t}} + R_{n}e^{j\Phi_{n}} \tag{4.17}$$

with

$$R_{t}e^{j\Phi_{t}} \approx \frac{1}{N^2} \sum_{\mathbf{x}\in\Omega_{\mathbf{W}}} \omega\left(\mathbf{x}\right) i^2\left(\mathbf{x}\right) r_{t}\left(\mathbf{x}\right) e^{j(\varphi_{t}(\mathbf{x})+\varphi_{n}(\mathbf{x}))},$$

$$R_{n}e^{j\Phi_{n}} \approx \frac{1}{N^2} \sum_{\mathbf{x}\in\Omega_{\mathbf{W}}} \left(1 - \omega\left(\mathbf{x}\right)\right) r_{n}\left(\mathbf{x}\right) e^{j\phi_{n}(\mathbf{x})},$$

$$\omega\left(\mathbf{x}\right) = \begin{cases} 1 & \text{if } \mathbf{x} \text{ at strong edges,} \\ 0 & \text{otherweise.} \end{cases}$$

In the case of weak image noise, the gradient distribution of strong edges still keeps a significant anisotropy. Hence, the first term of Equ. 4.17, $R_{t}e^{j\Phi_{t}}$, represents a smoothed version of strong edges. The second term, $R_{n}e^{j\Phi_{n}}$, can reduce noise by vector addition due to the large dispersion of noisy vectors. From this model it can be known that edges and noise are smoothed in the whole window region $\Omega_{\mathbf{W}}$. As a result, edges cannot be preserved by such isotropic filtering. Moreover, this model also indicates that edges and noise can be separated in local windows under proper assumptions. This property enables the design of anisotropic filters which should limit the smoothing within specified regions. The study of anisotropic filters is presented in the next section.

## 4.3 Edge-aware structure tensor

### 4.3.1 Design idea

In this section, the classic structure tensor is improved by means of tailoring local squared-gradients. Fig. 4.8(a) sketches a group of squared-gradients in a local window whose center is located at an edge. In this case, vectors similar to the one at the window center should take more votes for orientation estimation. In another situation, the window center may be a noisy pixel, as shown in Fig. 4.8(b). Noisy vectors are expected to be selected for the averaging. This concept is realized with an

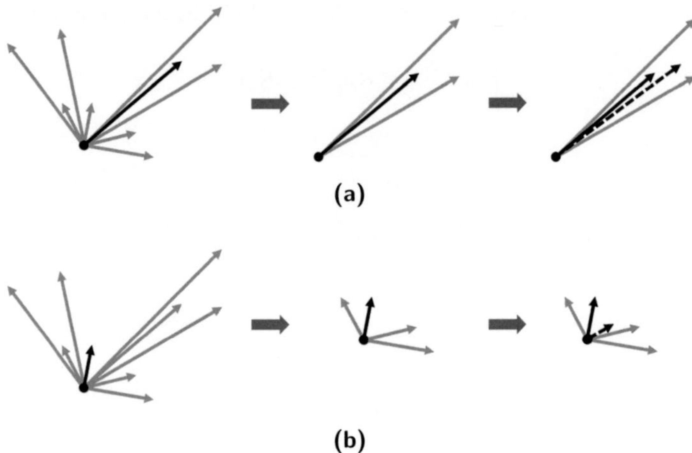

**Figure 4.8**  Improved orientation estimation (a) at an edge pixel and (b) at a noise pixel. Black solid lines are squared-gradients at window centers, while gray solid lines are squared-gradients in the neighborhood. In the first stage vectors with a norm significantly different from the window center are removed. In the second stage the mean squared-gradients are indicated with black dash lines.

algorithm that combines the classic structure tensor with a bilateral filter and an anisotropic filter. The achieved structure tensor is named as edge-aware structure tensor (EAST), which demonstrates superior characteristics in noise reduction and edge preservation. Large local windows are preferred in the structure tensor for significantly reducing noise. The proposed algorithm is able to reach the requirement on the structure selectivity depicted in Fig. 4.8.

### 4.3.2  Bilateral filter

The bilateral filter [51, 72] and its variants are important techniques in the world of edge-preserving filtering. This attractive image processing tool has been widely applied to image editing, image denoising and contrast enhancement. It takes the following form:

$$\check{g}(\mathbf{x}) = \frac{1}{C} \sum_{\mathbf{p} \in \Omega_W} g(\mathbf{p})\, G_W\left(\|\mathbf{p} - \mathbf{x}\|\right) G_R\left(|g(\mathbf{p}) - g(\mathbf{x})|\right), \qquad (4.18)$$

where $C$ is a factor for normalizing the sum. $G_W$ denotes a 2D Gaussian, which is related to the Euclidean distance between $\mathbf{x}$ and its neighbor $\mathbf{p}$. $\|\cdot\|$ means the $L_2$-norm. $G_R$ is a range function with a Gaussian form, which is attributed with $|g(\mathbf{p}) - g(\mathbf{x})|$, that is, the $L_1$-norm of image intensities. Since only similar intensities take part in the smoothing, strong edges can be preserved in filtered images. Furthermore, some authors [26, 54] improved the original bilateral filter with adaptive range functions relying on another image. This class of filters is known as the cross/joint bilateral filter, and takes a common form:

$$\check{g}(\mathbf{x}) = \frac{1}{C} \sum_{\mathbf{p} \in \Omega_W} g(\mathbf{p})\, G_W\left(\|\mathbf{p} - \mathbf{x}\|\right) G_R\left(|g_0(\mathbf{p}) - g_0(\mathbf{x})|\right). \qquad (4.19)$$

where $g_0$ is a reference image different from $g$.

With some extensions, the bilateral filtering can be integrated into the classic structure tensor. The new structure tensor is highlighted with the properties designed in Fig. 4.8. It is straightforward to extend the bilateral filter to vector fields or tensor fields [15, 81]. In these applications, the range filter, $G_R$, was linked with the distance of vectors or tensors. The filtering was carried out for each element of vectors or tensors. In Equ. 4.17 the mean squared-gradient was modeled by using a constant box kernel as the spatial filter. The local window was separated into edges and noise regions. If $\omega(\mathbf{x})$ is substituted with a bilateral kernel, the mean squared-gradient field can be expressed as

$$\mathbf{u}(\mathbf{x}) = \frac{1}{C} \sum_{\mathbf{p} \in \Omega_W} r(\mathbf{p})\, e^{j\varphi(\mathbf{p})} G_R\left(\|\nabla g(\mathbf{p})\| - \|\nabla g(\mathbf{x})\|\right). \qquad (4.20)$$

This mathematical expression can theoretically explain the principle of the bilateral structure tensor proposed in [81].

### 4.3.3 Adaptive filter kernels

The approach differs from the previous work in the improved structure selectivity, which leads to an anisotropic structure tensor. The filter kernel should be adaptively tuned in order to alleviate the structure mixture in local windows. In this work the Gabor-filter bank is used to detect multi-oriented structures. The Gabor-features, $\left\{ \tilde{g}_{\xi}\left(\mathbf{x}\right), \xi = 1, 2, \ldots, N_{\xi} \right\}$, extracted in Section 4.2.1.2 are capsulated into a vector,

$$\tilde{\mathbf{g}}\left(\mathbf{x}\right) = \left(\tilde{g}_1, \tilde{g}_2, \ldots, \tilde{g}_{N_{\xi}}\right)\left(\mathbf{x}\right),$$  (4.21)

because they belong to a same type of features. Additionally, the gradient magnitude, $\|\nabla g\left(\mathbf{x}\right)\|$, is also involved in the filter kernel. The filtering is expected to occur at strong edges. As a result, two range functions are used for the bilateral filtering. They are then combined in a single filter kernel:

$$G_{\sigma_1 \sigma_2}\left(\mathbf{p}, \mathbf{x}\right) = G_{\sigma_1}\left(\mathbf{p}, \mathbf{x}\right) G_{\sigma_2}\left(\mathbf{p}, \mathbf{x}\right)$$  (4.22)

with

$$G_{\sigma_1}\left(\mathbf{p}, \mathbf{x}\right) = \exp\left[ -\frac{\left(\|\nabla g(\mathbf{p})\| - \|\nabla g(\mathbf{x})\|\right)^2}{\sigma_1^2} \right],$$

$$G_{\sigma_2}\left(\mathbf{p}, \mathbf{x}\right) = \exp\left[ -\frac{\left\| \|\nabla g(\mathbf{p})\|^2 \tilde{\mathbf{g}}\left(\mathbf{p}\right) - \|\nabla g(\mathbf{x})\|^2 \tilde{\mathbf{g}}\left(\mathbf{x}\right) \right\|^2}{\sigma_2^2} \right].$$

Formally, the mean squared-gradient field is expressed as

$$\mathbf{u}\left(\mathbf{x}\right) = \frac{1}{C} \sum_{\mathbf{p} \in \Omega_{\mathbf{w}}} r\left(\mathbf{p}\right) e^{j\varphi(\mathbf{p})} G_{\sigma_1 \sigma_2}\left(\mathbf{p}, \mathbf{x}\right).$$  (4.23)

As explained in Section 4.2.2.3, the eigendecomposition of the structure tensor is equivalent to filtering a vector-valued image. In this sense, $G_{\sigma_1}$ and $G_{\sigma_2}$ are range filters for vector amplitudes and vector angles, respectively. $G_{\sigma_1}$ contributes to localizing strong edges, whereas $G_{\sigma_2}$ is an anisotropic filter responsible for searching oriented structures similar to the window center.

### 4.3.4 Tensor filtering

Let me include the structure tensor in the framework of bilateral filtering. The classic structure tensor (CST) defined in Equ. 4.7 can be rewritten as a weighted sum of local tensors, i.e.,

$$T_C(x) = \sum_{p \in \Omega_W} T(p) \, G_W(\|p - x\|) \tag{4.24}$$

with

$$T(p) = \begin{pmatrix} g_x^2 & g_x g_y \\ g_x g_y & g_x^2 \end{pmatrix}(p).$$

$G_W$ plays a role of the spatial filter. Furthermore, it is easy to derive Equ. 4.16 from Equ. 4.24 in that the eigendecomposition is a linear operation. In the same way, the edge-aware structure tensor (EAST) can be defined as

$$T_E(x) = \frac{1}{C} \sum_{p \in \Omega_W} T(p) \, G_{\sigma_1 \sigma_2}(p, x). \tag{4.25}$$

Similarly, the bilateral structure tensor (BST) can be expressed as

$$T_B(x) = \frac{1}{C} \sum_{p \in \Omega_W} T(p) \, G_{\sigma_1}(p, x). \tag{4.26}$$

### 4.3.5 Parameter selection

### 4.3.5.1 Parameters in Gabor filters

The performance of the EAST can be adjusted by a set of parameters. In the step of feature extraction, the number of orientations and scales, $N_\xi$ and $N_\eta$, as well as the lowest and highest center frequencies, $u_1$ and $u_2$, can fully determine the Gabor-filter bank. When the number of scales is larger than two, the remaining center frequencies at other scales can be automatically fixed according to the filter layout. Gabor filters are normally used to extract frequency components that correspond to meaningful image contents. However, the spatial frequency distributions of real images are often unknown. To deal with this problem, the Gabor-filter bank is constructed with following considerations. The number of scales, $N_\eta$, is set to 3. Furthermore, the lowest and highest normalized center frequencies are set to 0.125 pixel/cycle and 0.45 pixel/cycle, respectively. In this way, Gabor filters are configured to be oriented low-, band- and high-pass filters, which can cover the entire spectrum. The number of filter orientations decides the angular resolution for discriminating oriented structures. By setting $N_\zeta$ to 18, the angle interval of Gabor filters, $\Delta\theta$, is equal to $10°$.

### 4.3.5.2 Adaptive smoothing parameters

Both range filters used have a common Gaussian form with a zero mean. Thus, they can be fully determined by their standard deviations. Each range filter is observed as a sigmoidal threshold function which transforms feature distances into $[0,1]$. $G_{\sigma_1}$ serves as a detector of edge/non-edge structures. Its standard deviation plays a role of a soft threshold that classifies image structures in local windows according to the similarity of edge strength. In the EAST another range filter is used to deal with the problem induced by structure mixture. $G_{\sigma_2}$ classifies local orientations into two groups that are similar and dissimilar to the orientation of the window center. Since Gaussian noise does not indicate obvious orientations, noisy pixels in smooth regions are taken as similar in terms of the structure anisotropy. $\sigma_2$ plays a role of a soft threshold to discriminate the orientation similarity. The larger these two parameters are, the more strongly the squared-gradient vectors can be smoothed.

In most applications the smoothing parameter for the range filter is set to a globally unified value. The empirical studies reported in [82] manifest that the global parameter is not suited for detail preservation in all regions. The authors of [79] derived adaptive parameters from local phase characteristics. Their bilateral filter improved the perceptual effect of image denoising. [15] brought forward a statistical method for estimating the optimal smoothing parameters. However, this method relied on an iterative optimizing process which was not convenient to handle. In this work, an adaptive method for parameter estimation is developed. In Equ. 4.22 the range filters are attributed with

$$d_1\left(\mathbf{p}, \mathbf{x}\right) = \left| \|\nabla g(\mathbf{p})\| - \|\nabla g(\mathbf{x})\| \right|, \tag{4.27}$$

$$d_2\left(\mathbf{p}, \mathbf{x}\right) = \left\| \|\nabla g(\mathbf{p})\|^2 \tilde{\mathbf{g}}\left(\mathbf{p}\right) - \|\nabla g(\mathbf{x})\|^2 \tilde{\mathbf{g}}\left(\mathbf{x}\right) \right\|. \tag{4.28}$$

The smoothing parameters are associated with the distribution of feature distances in local windows. The effectiveness of the method is ascertained by the fact that the predented feature vectors are powerful for discriminating oriented structures.

Histogram-based approaches are normally used to study the distribution of statistical variables. To build the histogram of $d_k$ with $k = 1$ or 2, feature distances should be divided into bins. It is nontrivial to do this, because the dynamic range of $d_k$ varies in each local window. The size of bins will critically affect the histogram of $d_k$. Instead, the profile of sorted feature distances is studied in local windows. Let $\left\{\check{d}(l)\right\}$ be a set of increasingly sorted values of $d_k$ in a local window, where $l \in \{1, 2, 3, \ldots, l_{\max}\}$ denotes the index. For local windows of size $N \times N$, $l_{\max} = N^2 - 1$ holds. The feature distance located at the window center is ignored, because it is always equal to zero. Moreover, $\left\{\check{d}(l)\right\}$ implies the histogram of $d_k$, which is denoted as $H(\check{d})$ in this context. Note that each of range filters in the EAST is adopted to discriminate two sets of local structures. Edge and non-edge structures are separated in $G_{\sigma_1}$, while similarly and dissimilarly oriented structures are distinguished in $G_{\sigma_2}$. If image features for local structures have a high descriminative power, the histogram of $d_k$ will be unimodal or bimodal. Accordingly, the profile of $\check{d}(l)$ will have large convex or concave curvatures. Based on this relation, the distribution of $d_k$ can be separated into several partitions which represent obviously small, transitive and obviously large feature distances. Correspondingly, the shape of $\check{d}(l)$ is simplifed with connected line segments showing one or two inflection points. Fig. 4.9 schematically demonstrates four possible simplified profiles of $\check{d}(l)$, as well as the shapes of $H(\check{d})$ regardless of histogram bins. Furthermore, the indices underlying these line segments construct several sets containing classified pixels, which are labeled with $\mathbb{S}$, $\mathbb{A}$ and $\mathbb{D}$, respectively representing obviously small, transitive and obviously large feature distances mentioned above. $\mathbb{S} = \varnothing$ for type I, while $\mathbb{D} = \varnothing$ for type IV. By simplifying and splitting the profile of $\check{d}(l)$, a data-driven approach can be developed for parameter estimation.

For types I, III and IV, transitive structures lie in the tail of $H(\check{d})$. Thus, the structure classification is trivial in these cases. In comparison, the distribution of type II is dominated with transitive structures. To estimate $\sigma_1$, the data in $\mathbb{S}$ and $\mathbb{A}$ zones are taken into account. This ensures that the

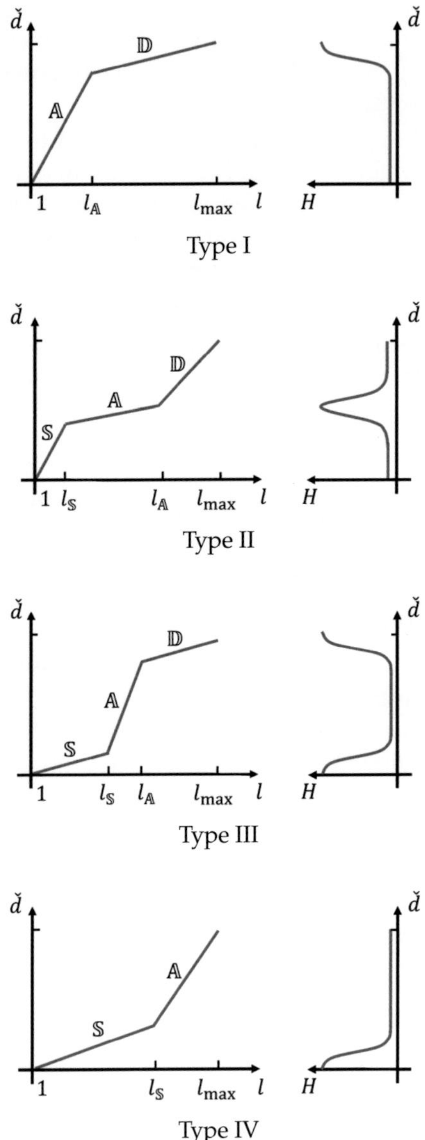


**Figure 4.9**  Left column: simplified profiles of feature distances. Right column: distribution models.

transitive regions from edges to smooth patches can be revealed in

$$\sigma_1 = \check{d}(1) + \sqrt{\frac{1}{l_{\mathbb{A}}} \sum_{l \in \mathbb{S} \cup \mathbb{A}} \left( \check{d}(l) - \check{d}(1) \right)^2}, \qquad (4.29)$$

where $\check{d}$ is derived from $d_1$, $l_{\mathbb{A}}$ is the upper limit of A zone. Considering the self-distance at the window center, the soft threshold is biased with $\check{d}(1)$. The second term of Equ. 4.29 is a rooted secondary center moment. Moreover, $\sigma_2$ implies an angular tolerance for detecting consistent orientations with respect to the window center. In the profiles of types I and II, the data in $\mathbb{S}$ and $\mathbb{A}$ zones indicate a wide range of angle differences. Therefore, these situations should be treated carefully. The estimate of $\sigma_2$ is $\check{d}(1)$ plus an empirical value that is smaller than the mean of sorted feature distances in $\mathbb{S}$ and $\mathbb{A}$ zones. Furthermore, the distributions of types III and IV have distinct peaks in the low value area of $\check{d}$. In these cases, only $\mathbb{S}$ zone is considered for estimating $\sigma_2$. $\mathbb{A}$ zone is not considered in order to ensure the accuracy of orientation selection. In summary, the estimation functions are formulated in terms of profile shapes:

for types I and II,

$$\sigma_2 = \check{d}(1) + \vartheta(\kappa), \qquad (4.30)$$

for types III and IV,

$$\sigma_2 = \check{d}(1) + \sqrt{\frac{1}{l_{\mathbb{S}}} \sum_{l \in \mathbb{S}} \left( \check{d}(l) - \check{d}(1) \right)^2}, \qquad (4.31)$$

where the values of $\check{d}$ are derived from $d_2$, $l_{\mathbb{S}}$ is the upper limit of $\mathbb{S}$ zone. Furthermore,

$$\vartheta(\kappa) = \frac{\kappa}{l_{\mathbb{A}}} \sum_{l \in \mathbb{S} \cup \mathbb{A}} \check{d}(l)$$

with $\kappa \in [0, 1]$. The choice of $\kappa$ is discussed later in Section 4.5.1. In a special case, the local window may contain constant intensities. $\sigma_1$ and $\sigma_2$ will be equal to zero. As this situation appears, $G_{\sigma_1 \sigma_2}$ is defined as a box kernel.

The last question is how to determine the inflection points in the simplified profile of $\check{d}(l)$. In the scheme, these points are calculated with the help of following indices:

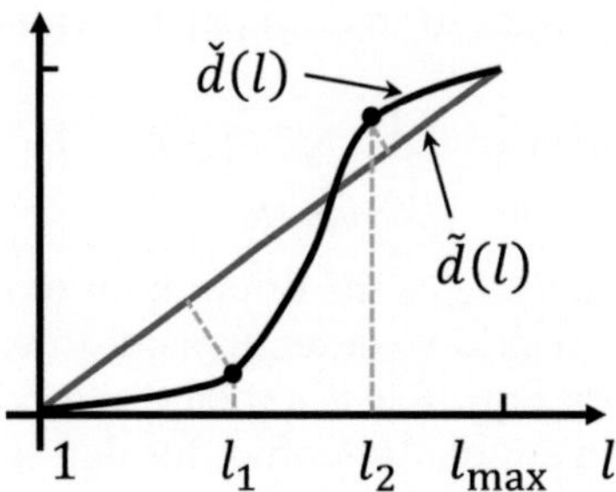


**Figure 4.10**  Estimating inflection points in $\check{d}$.

$$l_1 = \arg\min_l \left\{ \check{d}(l) - \tilde{d}(l) \right\},  \tag{4.32}$$

$$l_2 = \arg\max_l \left\{ \check{d}(l) - \tilde{d}(l) \right\},  \tag{4.33}$$

with

$$\tilde{d}(l) = \check{d}(l) + \frac{\check{d}\left(N^2 - 1\right) - \check{d}(1)}{N^2 - 1} l.$$

$\tilde{d}(l)$ denotes a straight line connecting the begin and end points of $\check{d}(l)$. The graphical illustration of $l_1$ and $l_2$ is shown in Fig. 4.10. $l_S$ and $l_A$ can be found according to Table 4.1.

**Table 4.1**  Upper boundaries of $S$ and $A$ zones.

| distribution types | $l_S$ | $l_A$ |
|---|---|---|
| I | 1 | $l_2$ |
| II | $l_2$ | $l_1$ |
| III | $l_1$ | $l_2$ |
| IV | $l_1$ | $l_{max}$ |

## 4.4 Detection scheme

### 4.4.1 Feature extraction

It has been known from Fig. 4.7 that orientations along straight edges show a unimodal distribution, whereas orientations along rough edges are randomly scattered. Moreover, multi-oriented structures like junctions possess two or more dominant orientations. Therefore, the orientation uniformity is high in individual modes, but low in the whole distribution. Smooth patches do not indicate an obvious anisotropy. It can be observed that plateau-honed surfaces consist of these four typical structures. Rough edges are regarded as defects. Since the orientation dispersion of noisy vectors makes no sense for the application, only the dispersion of edge orientations is investigated in the vector field.

$$\mathbf{u}\left(\mathbf{x}\right) = \lambda_1(\mathbf{x})\omega(\mathbf{x})e^{jO(\mathbf{x})}, \tag{4.34}$$

where $\omega(\mathbf{x})$ is the edge map created by segmenting the amplitude field, $\lambda_1(\mathbf{x})$, with Otsu's threshold [50]. Then, the orientation dispersion is measured with the small eigenvalue of the classic structure tensor. The window size is fixed to $5 \times 5$ in the assumption that such a small window can only contain a single edge. The derived small eigenvalues are denoted as $\check{\lambda}_2(\mathbf{x})$.

Since the edge strength is the main perceptual evidence in 2D gray value images to identify the severity of surface damage, image contrast is also an important cue for defect detection. Combining the contrast and orientation features of defective edges, the defect signature is ultimately defined as

$$S\left(\mathbf{x}\right) = \mathcal{N}\left(\lambda_1\right)\mathcal{N}\left(\check{\lambda}_2\right). \tag{4.35}$$

$\mathcal{N}\left(\cdot\right)$ is a min-max normalization function, which linearly transforms the maximum to one and the minimum to zero. Since eigenvalues have large dynamic ranges, $\lambda_1$ and $\check{\lambda}_2$ are logarithmically stretched before normalization. Because metal working is imperfect, it can be considered that defects always exist in surface samples. This means that slight orientation dispersion cannot be enhanced by normalization.

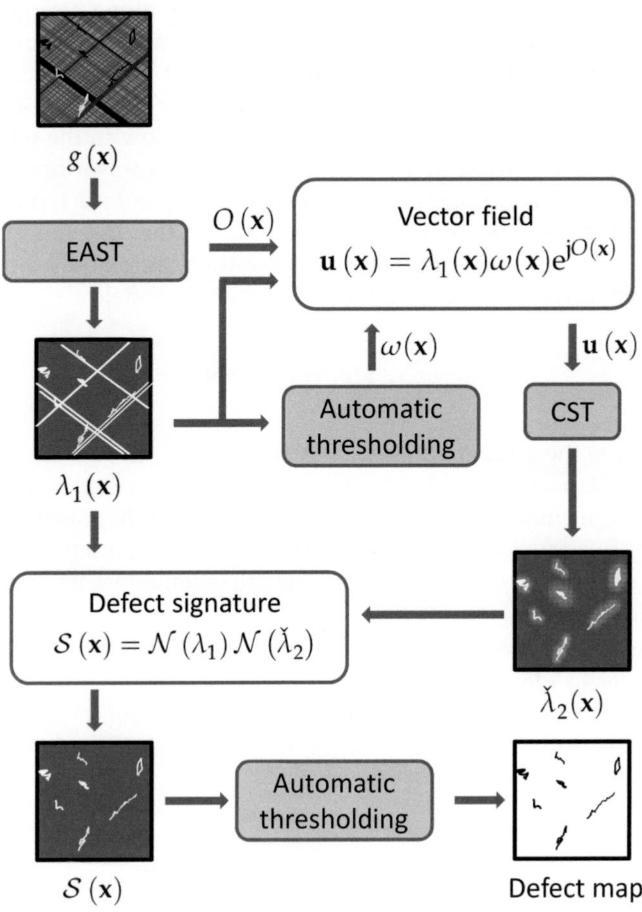



**Figure 4.11**  EAST-based detection scheme.

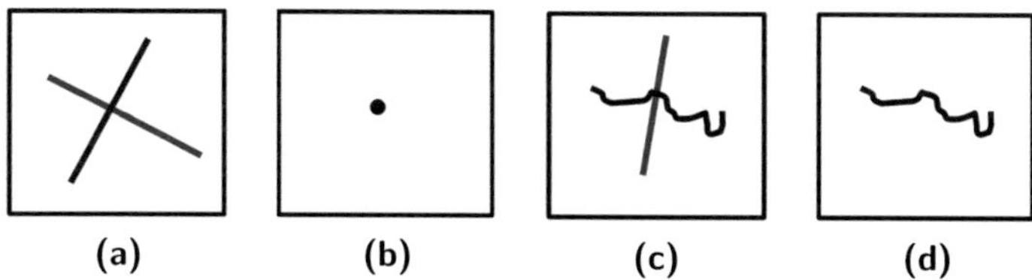

(a)       (b)       (c)       (d)

**Figure 4.12**  Detection at junctions. (a) Intersection of straight lines. (b) Detection result of (a). (c) Intersection of a rough line and a straight line. (d) Detection result of (c).

### 4.4.2 Segmentation

The most salient defects can be segmented by thresholding $\mathcal{S}(\mathbf{x})$ with Otsu's method. The EAST-based detection scheme is summerized in Fig. 4.11. Moreover, inhomogeneous intensities are irrelevant signals for surface characterization. These signals can be eliminated from the original image by preprocessing with homogenization techniques [4, 76]. Furthermore, it should be pointed out that any intersections at rough edges are a part of defects. In other words, only junctions of straight edges should be excluded from detection results. This consideration is explained with Fig. 4.12. In the postprocessing, isolated small and round regions are removed, since defective edges are large and elongated objects. The roundness measure is defined in Appendix C.1.

## 4.5  Experimental results

### 4.5.1  Range filters

In this section, filter kernels as well as parameter settings for the EAST are validated. The improvement of the filter kernel is illustrated by comparison with the BST. In the experiments, gradients are computed with horizontal and vertical central difference operators. Filter kernels are sized to $21 \times 21$ for both structure tensors. The smoothing parameter for the BST is automatically computed with Equ. 4.29. Moreover, the influence of $\kappa$ is investigated with an image pattern shown in Fig. 4.14(a). In this example, both the profiles of sorted $d_1$ and sorted $d_2$ are shaped as type I, as depicted in Fig. 4.13. Since edges are normally sparse in local windows, $\check{d}(l)$

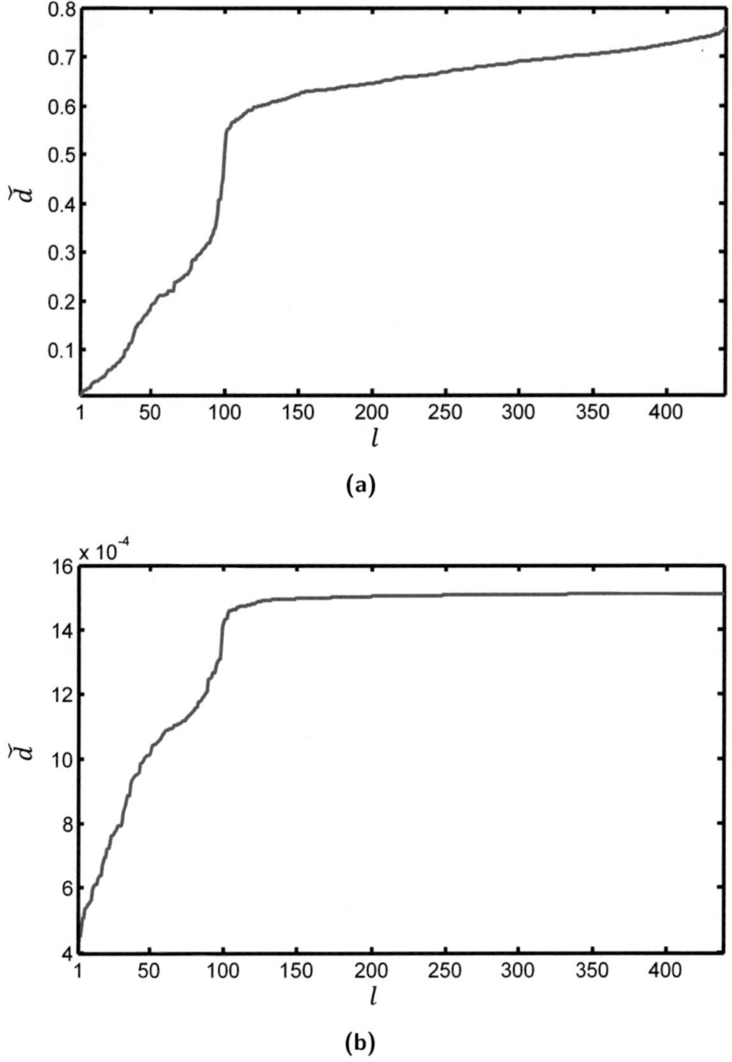

**Figure 4.13**  Profiles of $\breve{d}(l)$ obtained from (a) $d_1$ and (b) $d_2$. The local window is shown in Fig. 4.14.

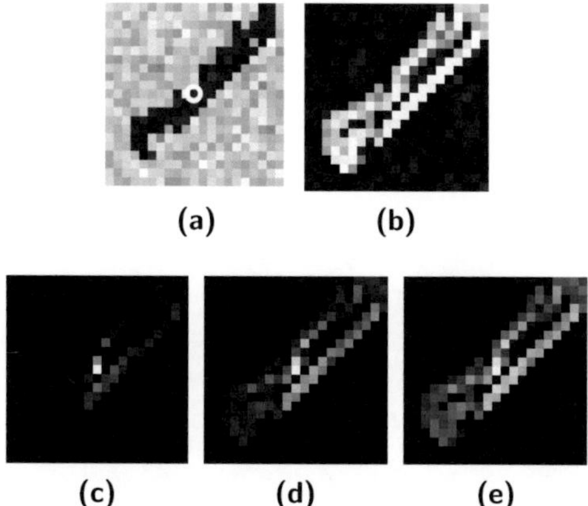

(a)        (b)

(c)        (d)        (e)

**Figure 4.14** Range filter adjusted by $\kappa$. (a) Local window containing a rough line. (b) $G_{\sigma_1}$, (c)-(e) $G_{\sigma_2}$ with $\kappa = 0.2, 0.6$ and $0.9$.

is likely to be types I or II at edges. Thus, the parameter, $\kappa$, is useful for adjusting the smoothing degree at edges. The window center marked with a white circle is located at a rough edge. With the estimated $\sigma_1$, the weights for edges are significantly higher than the surrounding in $G_{\sigma_1}$. This indicates that edges and non-edge regions can be reasonably classified by the automatic choice of $\sigma_1$. By observing Figs. 4.14(c)-(e), it can be noted that the larger the value of $\kappa$ is, the stronger the edge will be smoothed. In the application it is intended to preserve orientation fluctuations at rough edges. Thus, in the following $\kappa$ is set to 0.2 by default.

Fig. 4.15 demonstrates three synthetic patterns and corresponding adaptive filter kernels. In the first row the window center is located in a smooth region. In all three filter kernels, the weights for edges are obviously suppressed. This can effectively avoid the interaction between edges and smooth regions during the filtering. In this test, $\sigma_2$ shows its ability of discriminating oriented and non-oriented structures. $G_{\sigma_1}$ contributes to the localization of edges and smooth regions. Moreover, the superiority of the EAST can also be observed as edges are smoothed. As shown in the second row, $G_{\sigma_1\sigma_2}$ is oriented to the edge direction. Surrounding edges having similar directions also gain large weights in $G_{\sigma_1\sigma_2}$. In this case, dispersed orientations will not be processed together.

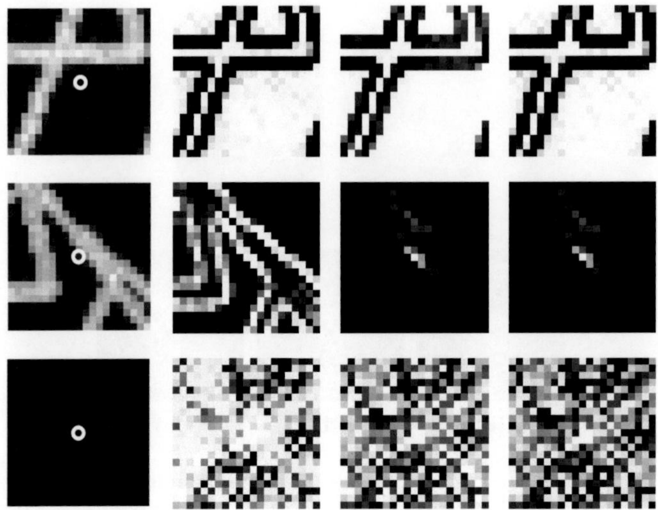

**Figure 4.15**   Range filters for artificial patterns. (a) Synthetic patterns, (b) $G_{\sigma_1}$, (c) $G_{\sigma_2}$, (d) $G_{\sigma_1\sigma_2}$.

These two experiments verify that the EAST exactly obeys the design idea demonstrated in Fig. 4.8. Besides, if a local window contains only noise, the BST and the EAST will attempt to find similar structures in noise. The example for this case is shown in the last row.

Furthermore, the BST and the EAST are tested by using real SEM images of plateau-honed surfaces. Practical examples for the typical structures mentioned in Section 4.2.2.3 are shown in Fig. 4.16. In complex technical surfaces, edges are likely to be intersected in a small area. Hence, $G_{\sigma_1}$ is not sufficient for generating accurate estimates of edge orientations. This problem can be effectively addressed with the anisotropic filter, $G_{\sigma_2}$. These examples also verify that the strategy for estimating smoothing parameters is applicable to both synthetic and real images.

### 4.5.2 Amplitude and orientation fields

In this section, the performance of the CST, the BST and the EAST is evaluated with an artificial picture. The artificial image, shown in Fig. 4.17, consists of random lines. Their positions and angles obey the uniform distribution. Since this picture is rich of junctions, it is especially suited

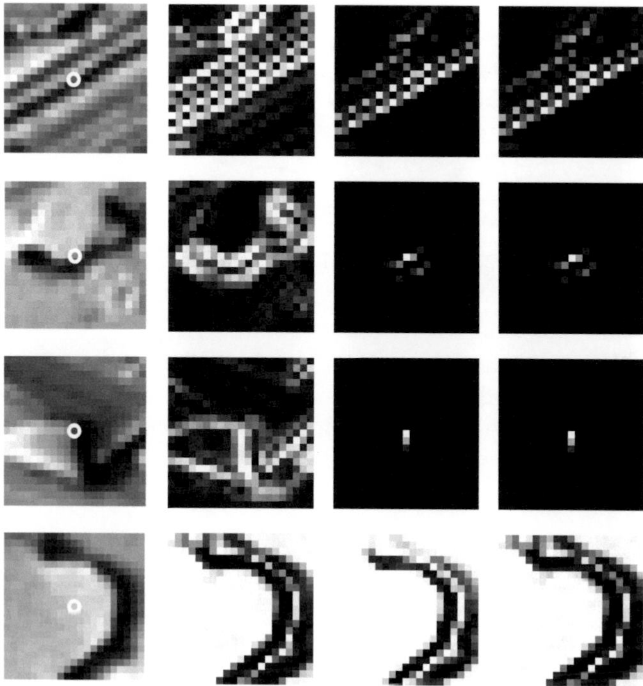

**Figure 4.16**  Range filters for real image patches. The rows from top to bottom are surface samples centered at a straight edge, a rough edge, a junction and a smooth patch. The columns indicate (a) Real image patches, (b) $G_{\sigma_1}$, (c) $G_{\sigma_2}$ and (d) $G_{\sigma_1\sigma_2}$.

for illustrating the advantage of using an anisotropic filter in the structure tensor. Moreover, the influence of parameters is also discussed. The CST and the BST are only affected by the window size, $N$, while the EAST has a set of parameters. As mentioned in Section 4.3.5, the performance of the EAST is mainly influenced by three parameters, $N_{\xi}$, $N$ and $\kappa$. During the visual evaluation, these parameters are firstly set to $\left(N_{\xi}, N, \kappa\right) = (18, 21, 0.2)$. Afterwards, one of these parameters is altered for the quantitative evaluation. To make objective assessments, a ground truth is needed in the experiments. Here the amplitudes of squared-gradients, $r(\mathbf{x})$, are taken as the ground truth of amplitudes. Besides, since the digital lines are not perfectly straight in a low resolution image,

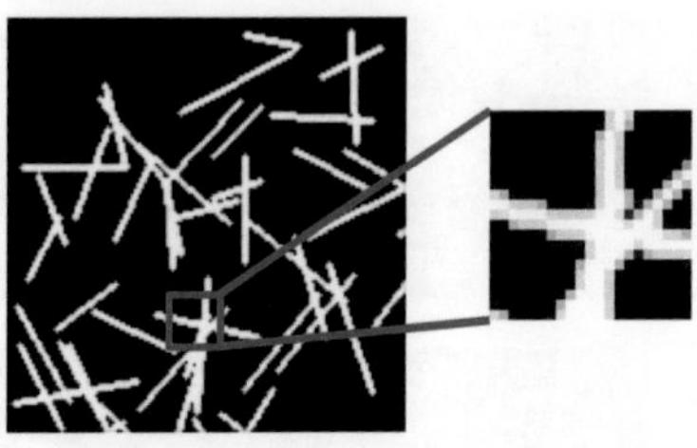

**Figure 4.17** Synthetic random lines.

the true edge orientations are assumed to be a slightly smoothed version of $\varphi(\mathbf{x})$. The CST with a window size of $3 \times 3$ is used for the smoothing. The ground truth of angles is denoted as $\breve{\varphi}(\mathbf{x})$, which is defined in $[-180°, 180°]$.

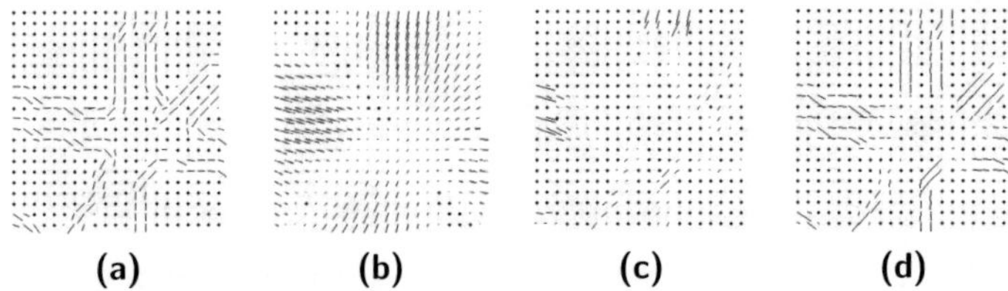

**(a)**        **(b)**        **(c)**        **(d)**

**Figure 4.18** Visual evaluation. (a) Ground truth. (b)-(d) Estimated vector fields associated with the CST, the BST and the EAST, respectively.

Figure 4.18 shows $r(\mathbf{x})\, e^{j(\breve{\varphi}(\mathbf{x})/2+\pi/2)}$ derived from noise-free gradients as well as $\lambda_1(\mathbf{x})\, e^{j(O(\mathbf{x})/2+\pi/2)}$ obtained by structure tensors. These vectors are chosen from a local window depicted in Fig. 4.17. In Fig. 4.18(b) the vectors obtained by the CST are strongly smoothed so that structure details become invisible. In Fig. 4.18(c), the vector amplitudes are almost zero at the junction. The drawback of the BST becomes clear in this example. In comparison, the EAST generates vectors that are most similar to the ground truth.

Furthermore, a quantitative assessment shall be conducted. Vector amplitudes and vector angles are evaluated separately. Vector amplitudes indicate the goodness of the edge localization, which can be described by the peak signal to noise ratio (PSNR):

$$\mathrm{PSNR} = 10 \log_{10} \frac{\max r^2\left(\mathbf{x}\right)}{\frac{1}{A} \sum_{\mathbf{x} \in \Omega_g} \left(\lambda_1\left(\mathbf{x}\right) - r\left(\mathbf{x}\right)\right)^2}, \tag{4.36}$$

where $A$ is the total number of pixels in the image. Vector angles reflect the orientation accuracy. The mean angular error (MAE) is computed with

$$\mathrm{MAE} = \frac{1}{2A_e} \sum_{\mathbf{x} \in \Omega_e} \mathrm{arc}\left(O\left(\mathbf{x}\right), \check{\varphi}\left(\mathbf{x}\right)\right), \tag{4.37}$$

$$\mathrm{arc}\left(O\left(\mathbf{x}\right), \check{\varphi}\left(\mathbf{x}\right)\right) = 180 - \left|180 - \left|O\left(\mathbf{x}\right) - \check{\varphi}\left(\mathbf{x}\right)\right|\right|,$$

where $\mathrm{arc}\left(O\left(\mathbf{x}\right), \check{\varphi}\left(\mathbf{x}\right)\right)$ [47] computes the smallest angular distance from $O\left(\mathbf{x}\right)$ to $\check{\varphi}\left(\mathbf{x}\right)$ in the unit circle. $O\left(\mathbf{x}\right) \in \left[-180°, 180°\right]$ is estimated by one of the structure tensors introduced in Section 4.3.4. The angles are selected from

$$\Omega_e = \left\{\mathbf{p}, r\left(\mathbf{p}\right) > 0\right\}.$$

$A_e$ is the size of $\Omega_e$. The mean orientation error is the half of the mean error of $O\left(\mathbf{x}\right)$.

In the following, it is illustrated how the number of orientations of the Gabor filter bank, $N_\zeta$, influences the performance of the EAST. In this test the angle resolution of the Gabor filter bank is altered from $5°$ to $30°$. Fig. 4.19 shows that the PSNR is approximately invariant to $N_\zeta$. This illustrates that the random lines shown in Fig. 4.17 are distinguishable when the angle resolution is less than $30°$. Thus, edge strength is not suppressed in the filtered amplitude field. However, the orientation accuracy is closely related to $N_\zeta$. This problem is illustrated in Fig. 4.20. If $N_\zeta$ is too small, multi-oriented edges cannot be exactly described with Gabor features. If $N_\zeta$ is too large, the anisotropic filter kernel of the EAST will be oriented to zigzag segments of discrete lines. In this experiment, the optimal orientation number for the Gabor-filter bank is found at $N_\zeta = 18$. As shown in Appendix A.1, this setting is also suitable for the filtering of other images.

Next, the parameter $\kappa$ is taken into account. In Fig. 4.21 the amplitude field is slightly degraded by an increasing $\kappa$. As known from Fig. 4.14, $\kappa$

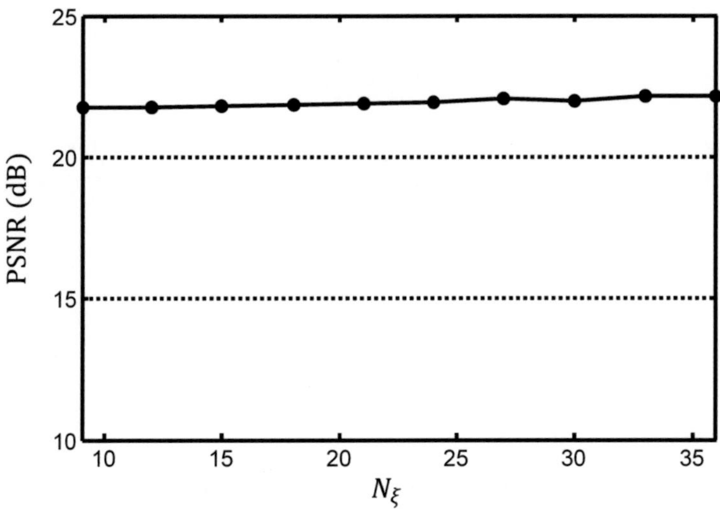

**Figure 4.19**  Amplitude evaluation of the EAST depending on $N_\xi$.

does not affect the edge localization. Therefore, the degradation is not induced by the blurring across edges, like in Fig. 4.18(b), but by the smoothing of junctions, like in Fig. 4.18(c). Moreover, the MAE curve is shown in Fig. 4.22. Choosing a value of $\kappa$ that is too small will result in an undersmoothing of squared-gradients. Angle errors will be large due to noise. In addition, if the value of $\kappa$ is too large, the anisotropy of $G_{\sigma_2}$ will be weakened. In this case, the orientation accuracy is lowered by the interaction of multi-oriented structures. Over-smoothed orientations cannot exactly reflect the shape of edges.

The third test is to investigate the influence of the window size, which is a common parameter for the CST, the BST and the EAST. Therefore, a comparative study is performed. Test results are exhibited in Fig. 4.23 and Fig. 4.24. The CST is most sensitive to the window size. Poor edge localization causes strong amplitude biases at both edge and non-edge locations. The BST is better than the CST, since for the BST the filtering does not take place across edges. In comparison, the EAST shows the best edge-preserving ability. The estimated amplitude field is nearly the same whatever the window size is. The angle accuracy of the CST and the BST depends on the degree of structure mixture in local windows.

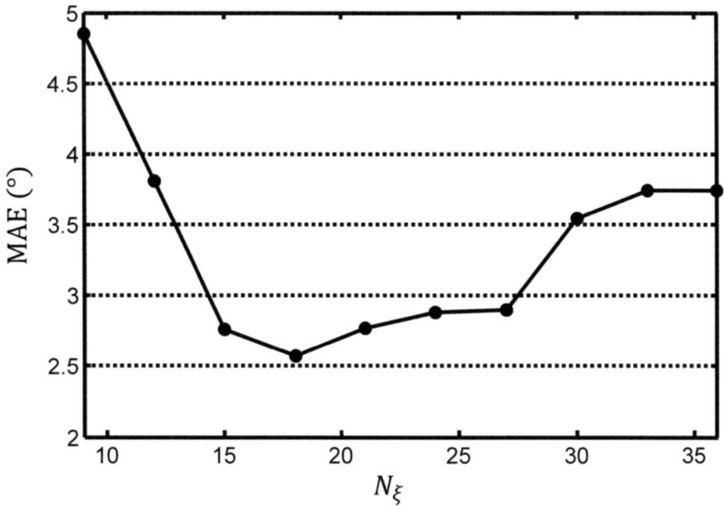

**Figure 4.20** Angle evaluation of the EAST depending on $N_\xi$.

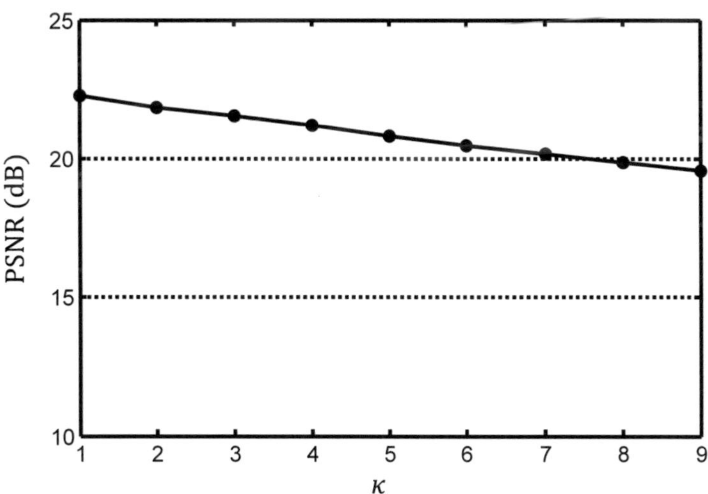

**Figure 4.21** Angle evaluation of the EAST depending on $\kappa$.

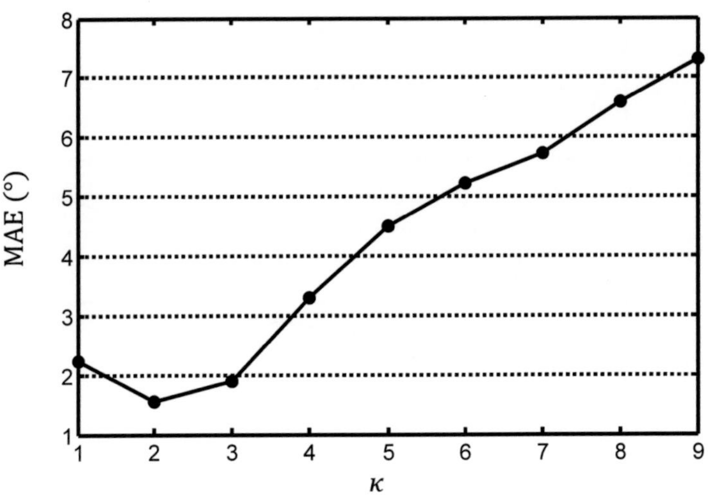

**Figure 4.22**   Angle evaluation of the EAST depending on $\kappa$.

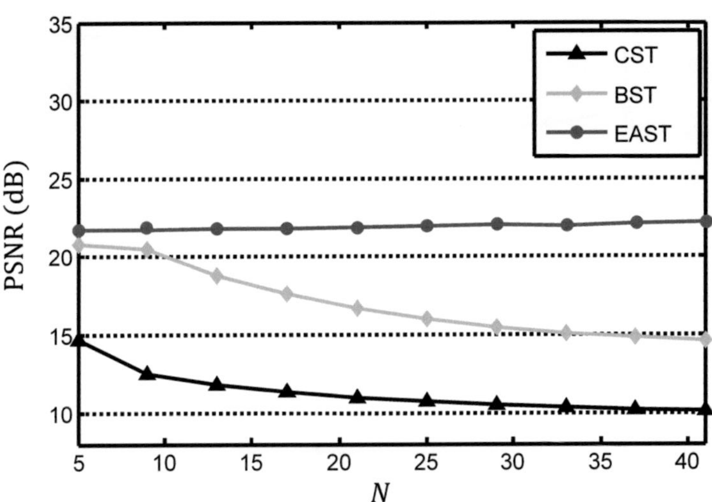

**Figure 4.23**   Amplitude evaluation of structure tensors depending on $N$.

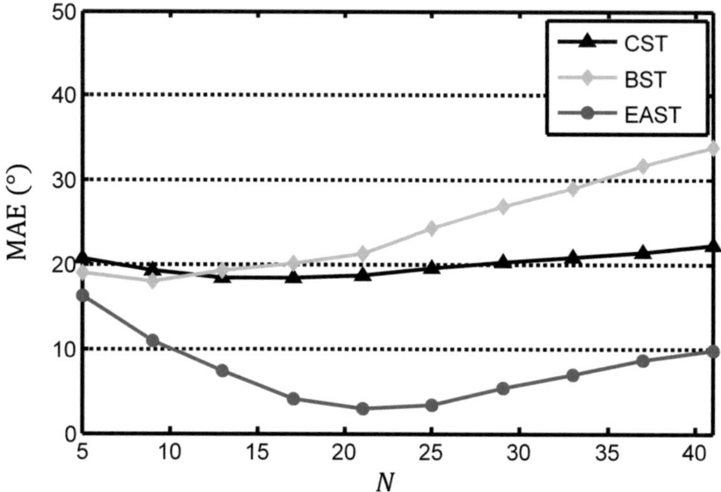

**Figure 4.24** Angle evaluation of structure tensors depending on $N$.

Hence, large windows are not suitable for estimating edge orientations. The EAST breaks through this limitation and achieves good estimates for angles. Local windows used by the EAST specify the area for searching similar structures. The larger the window becomes, the more structures that resemble the window center are likely to be found. Consequently, the squared-gradient field will be strongly filtered by using large windows. Hence, the under- and over-smoothing problems persist as well in the selection of the window size. In this test the optimal window size is $N = 21$.

Lastly, the noise resistance of all three structure tensors is illustrated. Gaussian noise with different standard deviations is added to Fig. 4.17. The noise level, $\sigma_n$, is denoted as the ratio of the noise standard deviation to the maximum gray value of the image. Once multi-oriented structures cannot be separated in the filter kernel, amplitudes at junctions will be blurred or weakened. PSNR values will be thus decreased. Fig. 4.25 shows the PSNR curves computed for the CST, the BST and the EAST. As the noise level is increased, PSNR values for the EAST rapidly declines from a large value. This result can be interpreted by recalling the vector model presented in Equ. 4.17, i.e.,

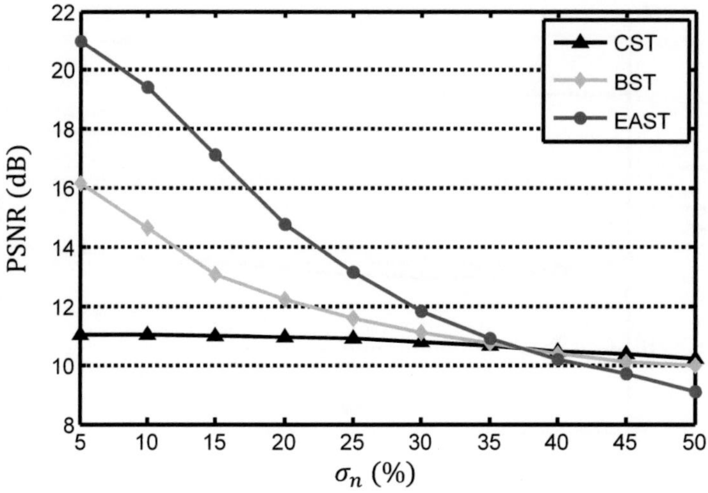

**Figure 4.25**   Amplitude evaluation of structure tensors depending on $\sigma_n$.

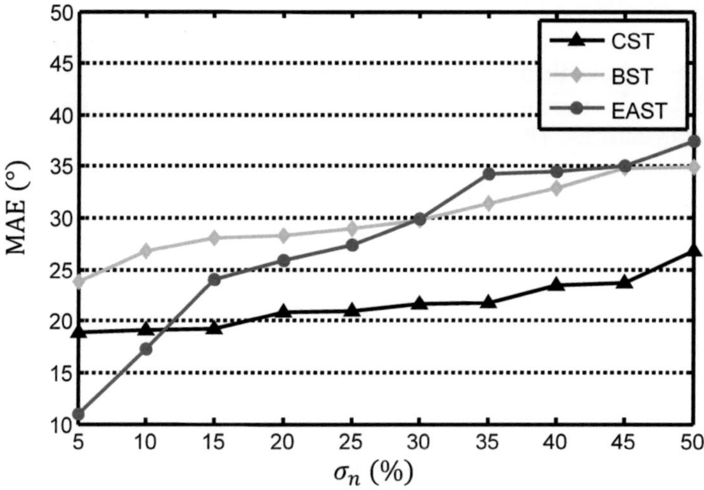

**Figure 4.26**   Angle evaluation of structure tensors depending on $\sigma_n$.

$$R_{\mathbf{W}}e^{j\Phi_{\mathbf{W}}} = R_t e^{j\Phi_t} + R_n e^{j\Phi_n}.$$

In this model, additive noise is split into two parts. The one is explicitly represented by $R_n e^{j\Phi_n}$, the other is included in $R_t e^{j\Phi_t}$. The range filter, $G_{\sigma_1}$, used in the BST and the EAST, intends to remove $R_n e^{j\Phi_n}$ and preserve $R_t e^{j\Phi_t}$. In the case where weak noise is superposed on strong edges, the local orientation can be accurately estimated with $R_t e^{j\Phi_t}$. However, this assumption will be violated when the image is contaminated by strong noise. If edges are not perceptible in strong noise, the EAST and the BST will attempt to preserve noise edges. In contrast, the CST blurs all edges in the amplitude field. This leads to strong noise reduction but poor edge localization. Moreover, MAE curves shown in Fig. 4.26 also illustrate the same issue. Angular errors generated by the EAST are small when image noise is weak and are then quickly increased as the noise level becomes high. For the application, current SEMs can create high quality pictures. Image noise is not critical for inspecting cylinder bore surfaces.

### 4.5.3 Applications

In this section the usefulness of the EAST for surface inspection is evaluated. The SEM image shown in Fig. 3.6 is adopted for the test. Due to the good image quality, image preprocessing is not needed in the experiments. Parameters are also set to $\left( N_{\xi}, N, \kappa \right) = (18, 21, 0.2)$. The process of EAST-based feature extraction is exhibited in Fig. 4.27. By observing these feature maps, it can be seen that noise and straight edges are gradually removed. Defects are highlighted in the defect signature map. Instead of the EAST, the CST and the BST are also applied to the calculation of the signature of the defect. These two variants of the detection method are involved in the following tests.

Then, global and local detection approaches are to be quantitatively evaluated. Since the real ground truth is not available for complex technical surfaces, defective edges and salient groove edges are manually marked in the original image. The marker is shaped as a small window of size $2 \times 2$. Figure 4.28 demonstrates the achieved map of markers. Furthermore, an evaluation method for error analysis is visualized in Fig. 4.29. False positive errors (FP) correspond to false alarms at groove edges. Missed defects are regarded as false negative errors (FN). True

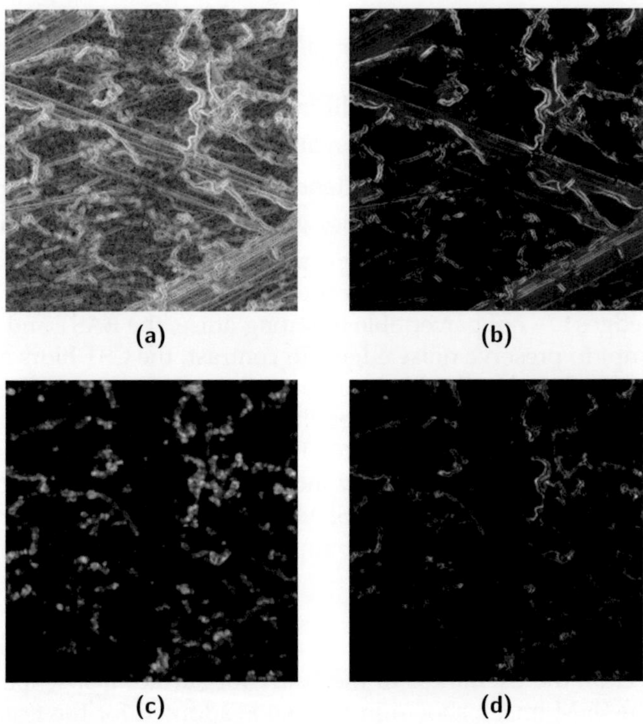

**Figure 4.27** Feature extraction. (a) —g—, (b) $\lambda_1$, (c) $\check{\lambda}_2$, (d) $\mathcal{S}$.

positive detections (TP) are the correct segmentations of defects. Marked groove edges that are not detected as defects represent true negative detections (TN). All these measures are computed taking the number of marked pixels into account. Based on these notions, the true positive rate, $P$ (Defect|Defect), and the false positive rate $P$ (Defect|Groove), are calculated as follows:

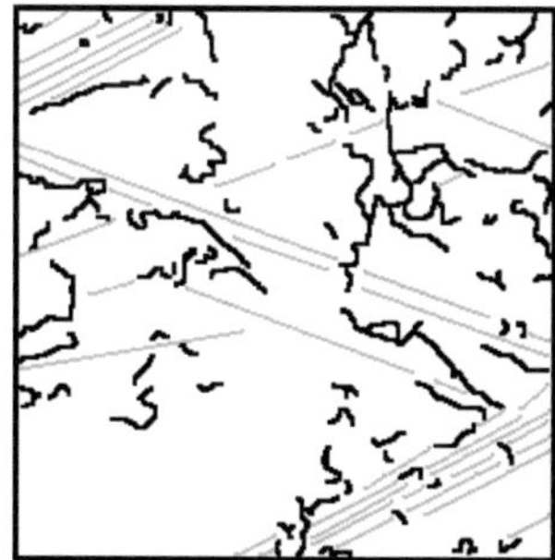

**Figure 4.28** Ground truth for the SEM test image. Black markers are defective edges, while gray markers are groove edges.

**Figure 4.29** Measures for error analysis.

$$P\left(\text{Defect}|\text{Defect}\right) = \frac{\text{TP}}{\text{TP} + \text{FN}}, \tag{4.38}$$

$$P\left(\text{Defect}|\text{Groove}\right) = \frac{\text{FP}}{\text{TP} + \text{FP}}. \tag{4.39}$$

The descriptive power of the defect signature can be verified with receiver operating characteristic (ROC) curves [71]. A ROC curve is created by relating the true positive rate with the false positive rate at different threshold settings. Figure 4.30 shows ROC curves for the EAST-based method, where different values of $\kappa$ and different window sizes are used. These curves do almost overlap. This indicates that the performance of the EAST-based method is not significantly influenced by these parameters. In contrast, the CST- and the BST-based methods are instable when the window size is changed. This issue becomes obvious in Fig. 4.31 and

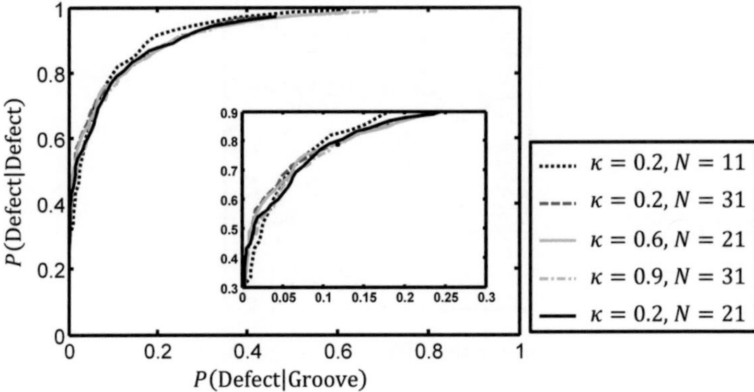

**Figure 4.30**  Evaluating the EAST-based approach in ROC curves.

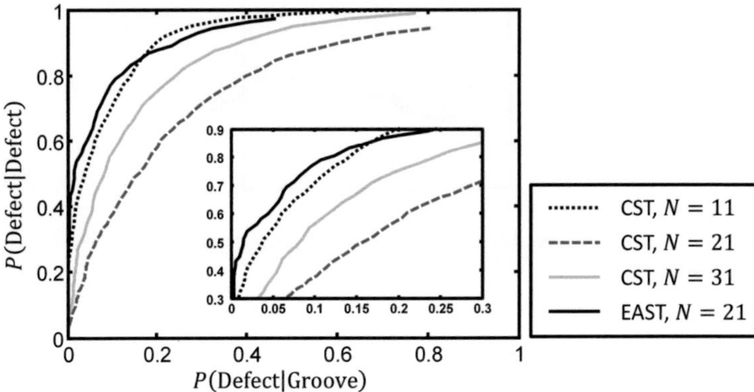

**Figure 4.31**  Evaluating the CST-based approach in ROC curves.

Fig. 4.32. In addition, these three methods show a similar performance as the window size is small. Since in small windows the risk for a structure mixture can be greatly reduced, the benefit of using the anisotropic filter is not obvious. It should be pointed out that small windows are unable to detect orientation variations at large scales. Small windows will also lower the noise immunity of local approaches. Therefore, large windows are preferred in the EAST-based method.

The accuracy of segmention is not emphasized in the evaluation that

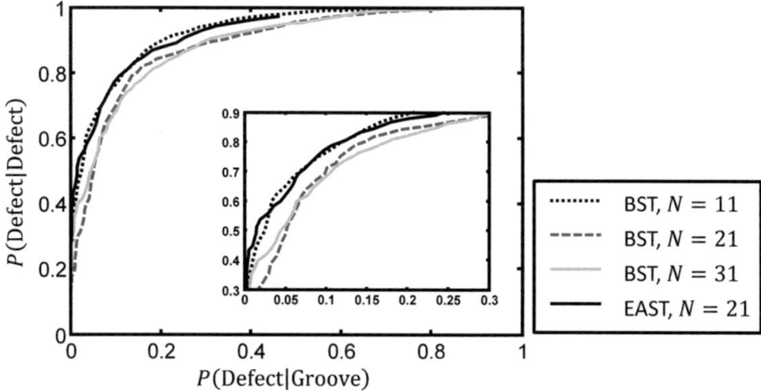

**Figure 4.32** Evaluating the BST-based approach in ROC curves.

uses a ground truth. The false positive errors are further evaluated with a set of comparative studies. Figure 4.33 provides a comprehensive illustration of the detection accuracy. The condition probabilities $(P(\text{Defect}|\text{Defect}), P(\text{Defect}|\text{Groove}))$ are given under the images. In this way, detection results at different parameter settings are associated with numerical evaluation results. In the output images, defective locations are marked in white. It can be seen that both the true positive rate and the false positive rate are improved as the value of $\kappa$ and the window size $N$ are increased. Simultaneously, the localization of defects becomes slightly worse. Nevertheless, the detection results are quite similar. From this test it can be known that the overall performance of the EAST-based detection scheme is not critically affected by the parameter settings. In Fig. 4.34 detection results generated by the CST-based method are placed together. As the window size is increased, both the detection rate and the segmentation results are degraded. A false alarm rate of more than 10% is generally unacceptable for practical applications. The BST-based method achieves a higher relative true positive rate; see Fig. 4.35. However, the false alarm rate is still too high. Moreover, both CST- and BST-based methods yield high true positive rate at the expense of a poor localization. Hence, it can be argued that these two methods are unstable with respect to parameter settings. In comparison, the global detection scheme cannot achieve satisfactory result, as shown in Fig. 4.36, when the surface quality is seriously degraded.

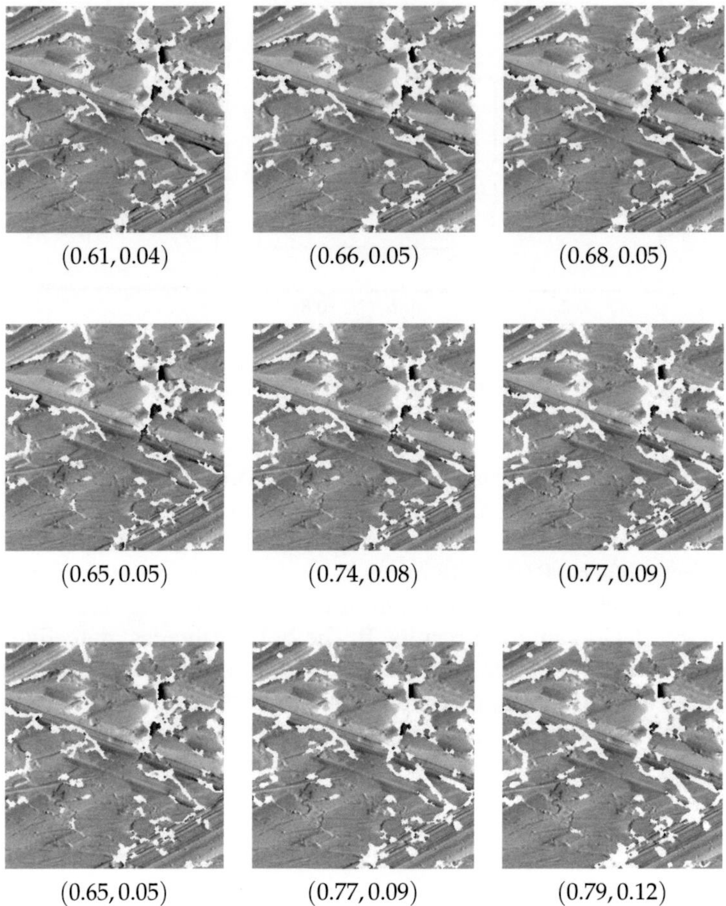

**Figure 4.33**  Evaluating EAST-based segmentation results. Columns: from left to right $\kappa = 0.2, 0.6, 0.9$. Rows: from top to bottom $N = 11, 21, 31$.

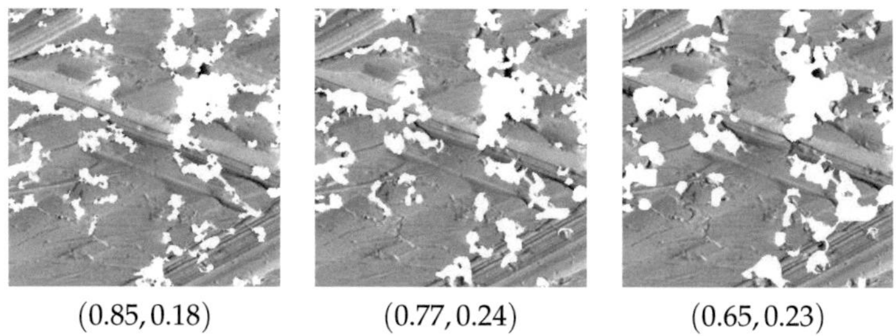



**Figure 4.34**   Evaluating CST-based segmentation results. From left to right $N = 11, 21, 31$.

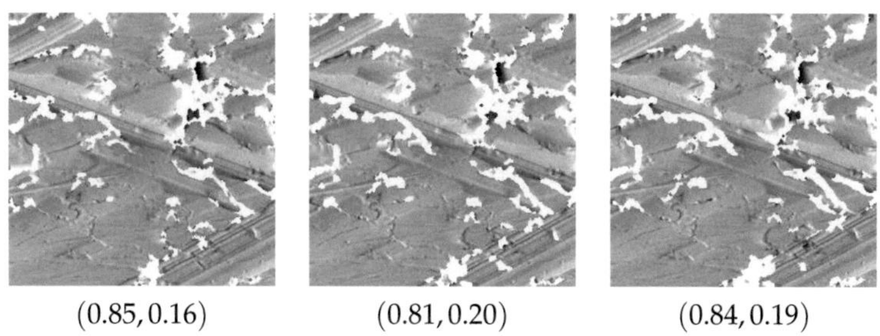



**Figure 4.35**   Evaluating BST-based segmentation results. From left to right $N = 11, 21, 31$.

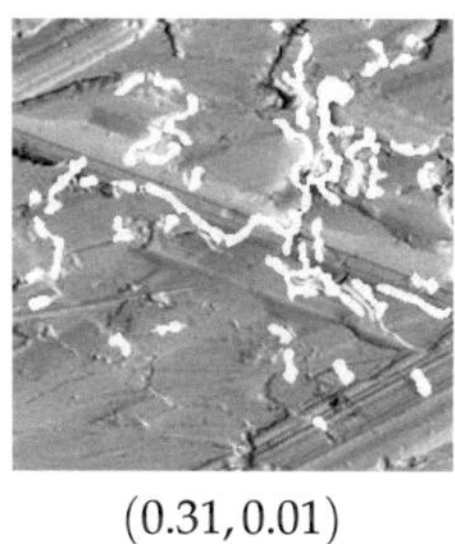



**Figure 4.36**   Evaluating the global method.

# 5 Graphite detection

The mechanical properties of engine cylinders can be optimized by incorporating vermicular graphite in cast iron. Exposed graphite particles could fall off after test running. The empty cavities on cylinder bore surfaces result in porous structures which constitute a micro-pressure chamber system. The multi-stage honing process is not sufficient to uncover graphite grains. A further treatment with laser irradiation is needed to ablate a very thin layer of metal from bore surfaces. After the laser exposure, metal folds are removed. Worm-like graphite particles are expected to become visible throughout the surface. Moreover, deep honing grooves may remain on the surface. They represent oil channels connecting pores, and are undesired in the surface finish. Figure 3.2 shows a series-production cylinder bore surface. Textural disturbances and non-uniform intensities are two major problems for detecting graphite grains. In this chapter an inspection algorithm is designed for segmenting graphite particles on laser-exposed cylinder bore surfaces. The approach builds upon a new understanding of how to describe the image topography in morphological scale spaces.

## 5.1 Related work

The metallographic characteristics of graphite cast irons have been intensively studied in material science. The mechanical properties of work pieces are very dependent on the proportion of spheroidal, vermicular and lamellar graphite in cast iron. Several image analysis algorithms for graphite classification [30, 49, 57] were proposed. These algorithms quantitatively describe the shape and distribution of uncovered graphite particles. The method presented in this chapter is comparable with them in the binarization of specimen images.

Local adaptive thresholds are popularly used to segment graphite grains in LOM images. These techniques were specially developed to cope with the intensity non-uniformity. The approaches for computing the threshold locally have been summarized in a survey paper [70]. The representative methods are Niblack's, Sauvola's and Bernsen's thresholds, which are parameterized with different inhomogeneity measures. The segmentation criteria proposed by Niblack and Sauvola tuned threshold values with the local mean and the local variance in different forms. Bernsen chose the mean of the local maximum and minimum as the threshold if the local contrast was large enough. The thresholding techniques mentioned above are effective for well-prepared specimens of graphite cast irons. In the case where surface textures are removed by polishing, a good image contrast can be easily obtained by using a bright field illumination. The segmentation of dark graphite grains and the bright background was tractable. However, graphite inspection on laser-exposed surfaces may suffer from strong textural disturbances, which result in poor segmentation results when adaptive thresholds are used. This problem is addressed with a novel strategy based on the image topography. The discrimination between foreground objects and background textures will be accomplished pixelwise.

In this work gray-level images are interpreted as topographic maps by taking the image intensity as the third extent besides the spatial coordinates. In this sense, the concept behind the approach is similar to the idea of analyzing the 2.5D surface topography. According to the literature, mathematical morphology turns out to be a powerful tool for surface characterization [60]. It can be found that the results achieved by Decencire et al. [12] are quite close to presented approach, although their approach was originally designed for the measurement of surface roughness. With an alternate sequential filter (ASF) [53], they decomposed the surface topography into three elements — reference surface, superficial roughness and valleys. Analogously, the approach separates the topography of LOM images into different roughness levels. Graphite grains are to be detected in the highest roughness level.


## 5.2 Morphological scale spaces

In this section a method is proposed for characterizing the image topography by multiscale morphological filtering. This section starts with the notation of the morphological operators used in this context. By analyzing the scale-space behaviors of hat transforms, topographical shapes can be reasonably interpreted with feature curves. Numerical shape signatures are then derived by dual observations of morphological scale spaces. The application to the graphite detection will be detailed in Section 5.2.1.

### 5.2.1 Notation

Mathematical morphology [39] is a nonlinear methodology for image analysis. The basic morphological operations are dilation and erosion. Sequential calculation of dilation and erosion leads to the morphological opening and closing. Considering a signal $g(l)$, $l \in \Omega_g$, to be a function mapping a linear coordinate into $\mathbb{R} \cup (-\infty, +\infty)$, the dilation of $g(l)$ by a flat structuring element $B(l)$ can be denoted as:

$$(g \oplus B)(l) = \bigvee_{l \in \Omega_g} \{g(l_x) + B(l - l_x)\}, \tag{5.1}$$

where $\vee$ is a supremum operator. $l_x$ denotes the linear coordinate in the neighborhood of $l$. Accordingly, the erosion is defined as:

$$(g \ominus B)(l) = \bigwedge_{l \in \Omega_g} \{g(l) + B(l_x - l)\}, \tag{5.2}$$

where $\wedge$ is an infimum operator. With the definition of dilation and erosion, the opening can be expressed as:

$$y_o(l) = (g \circ B)(l) = (g \ominus B \oplus B)(l). \tag{5.3}$$

Then, the closing can be written as follow:

$$y_c(l) = (g \bullet B)(l) = (g \oplus B \ominus B)(l). \tag{5.4}$$

Furthermore, the top-hat transform is achieved by subtracting the opened image from the original one,

$$h(l) = (g - g \circ B)(l), \tag{5.5}$$

and calculate the bottom-hat transform by subtracting the original image from the closed one (in order to obtain a positive result),

$$\check{h}(l) = (g \bullet B - g)(l). \tag{5.6}$$

Note that only the peaks (valleys) smaller than the structuring element can be detected by the top-hat (bottom-hat) transform. Therefore, the multiscale top-hat and bottom-hat transforms can be formulated by incorporating the size of structuring elements as an additional attribute:

$$h(l,s) = (g - g \circ B_s)(l), \tag{5.7}$$

$$\check{h}(l,s) = (g \bullet B_s - g)(l), \tag{5.8}$$

where $s$ is a natural number representing the scale index of a structuring element. $B_s(l)$ is computed by recursively dilating $B(l)$ with itself:

$$B_s(l) = \left( \underbrace{B \oplus B \oplus \cdots \oplus B}_{s \text{ times}} \right)(l). \tag{5.9}$$

In particular, $h(l,0)$ and $\check{h}(l,0)$ are designated as zero, i.e.,

$$h(l,0) = \check{h}(l,0) \equiv 0. \tag{5.10}$$

In the topographic analysis, the top-hat transform was used to extract bright features from peaks, while the bottom-hat transform was utilized to locate dark features at valleys [44].

### 5.2.2 Scale-space behavior

To facilitate the visualization, an example is shown in Fig. 5.1(a). A discrete 1D signal is opened with a series of line-like structuring elements. The length of the structuring element was sequentially selected from the set $\{3, 5, 7, \ldots, 21\}$, i.e., the scale index, $s$, ranged from 1 to 10. The original signal and opened signals were superposed in a single diagram. The opening operation flattened the signal by cutting down its peaks. Thus, the opened signals can be viewed as cutting lines. Analogously, the closing operation smoothed the signal by filling up its valleys. Due to the duality of both hat transforms similar properties are obtained from the bottom-hat scale space. In the following, the discussions concentrate on the top-hat scale space derived by opening.

By observing a pixel position, the traced path through the top-hat scale space creates a bright feature curve. The amplitude increment of peaks is equal to the falling distance of the cutting line. Thereby, the original signal can be decomposed into several lattices. For a specified scale range larger than the scale $s - 1$ and smaller than the scale $s$, the extracted lattice represents the bright feature. Formally, the bright feature can be computed with

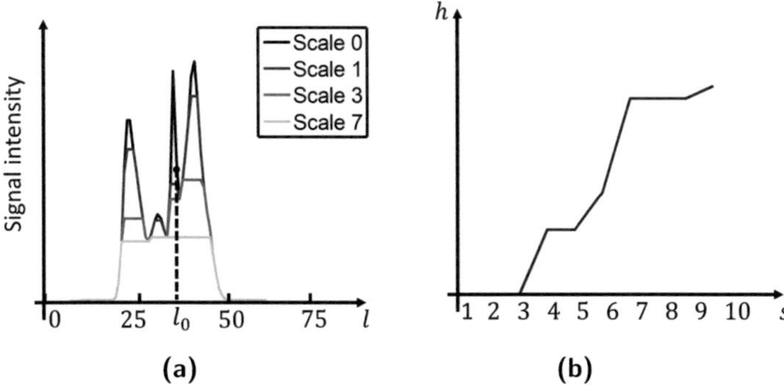

(a)  (b)

**Figure 5.1**  Scale-space behaviors. (a) Opened signals. (b) Bright features at $l_0$.

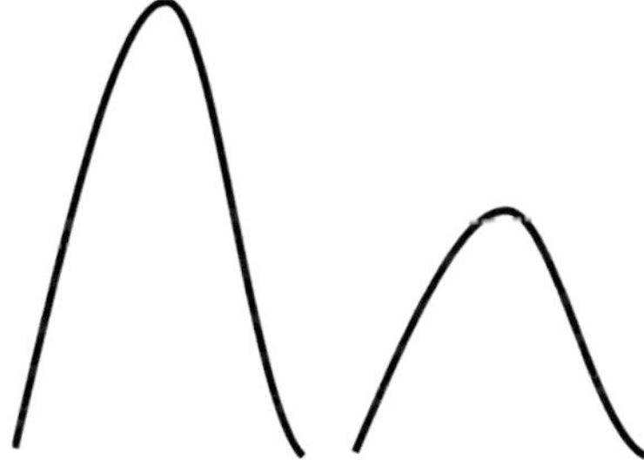

**Figure 5.2**  Peak sharpness influenced by the height and the width.

$$\Delta h(l,s) = h(l,s) - h(l,s-1), \tag{5.11}$$

In a similar way, the dark feature is defined as

$$\Delta \check{h}(l,s) = \check{h}(l,s) - \check{h}(l,s-1). \tag{5.12}$$

It should be noted that bright and dark features are not always available at signal points because they are scale-limited. This issue can be illustrated by observing a signal point at $l_0$. From Fig. 5.1(b) it can be seen that no bright features are obtained from the top-hat transform at low scales $(1{\sim}3)$, since the peak that covers $l_0$ is broader than 7 points (equivalent to the size of the structuring element at scale 3). Thus, the cutting line coincides with the original signal at $l_0$. By enlarging the size of the structuring element, the peak between scales $(3\sim4)$ is detected until the cutting line dropped off to a local minimum and then stops to change. The falling process of the cutting line continues when the size of structuring element reaches the width of a larger peak covering $l_0$. Thus, the magnitude of the bright feature curve does not increase in the scale interval $(4\sim5)$. The same situation happens at scales $(7\sim9)$. This scale-space behavior results in a stair-like bright feature curve. Furthermore, It should be pointed out that the slope of bright feature curve is related to the sharpness of the peaks. Figure 5.2 shows this property intuitively. The high peak is sharper than the low one as two peaks have the same bottom width (at a fixed scale). In other words, when extracting bright features from two neighboring scales, the thickest lattice (the largest slope) of bright features will obtained from the highest peak. The analog properties can also be obtained in the bottom-hat scale space. Combining both top-hat and bottom-hat scale spaces enables a comprehensive description of signal roughness.

### 5.2.3 Morphological stability

The characterization of image topography is normally related to observation scales. For instance, small valleys may belong to large peaks and thus cannot be detected in large scales. In turn, small peaks located at large valleys will be neglected when large valleys are in the focus. Hence, pixels may change their roles in the image topography with regard to the observation scale. This phenomenon often appears on complex technical surfaces, and can be explained with the fractal theory [16]. However, some valleys or peaks are rarely confused with their counterpart, when their size or intensity stands out in the neighborhood. The fractal characteristic is destroyed at these places. In optical micrographs graphite grains are always shown as isolated dark particles. They represent significant image valleys, even if rough groove textures are distributed in their vicinity. Utilizing this feature graphite grains can detected. For the intuition, topographical shapes are graphically illustrated in Fig. 5.3. By enlarging the observation domain from (a) to (b), the morphological shape varies from a peak to a valley. From this example it can be known that visual recognition is accomplished by dual observations of topographical maps, from above and below, respectively. The visual classification of topographical shapes can be formulated with the definitions given below. $l_0$ represents an observed signal point.

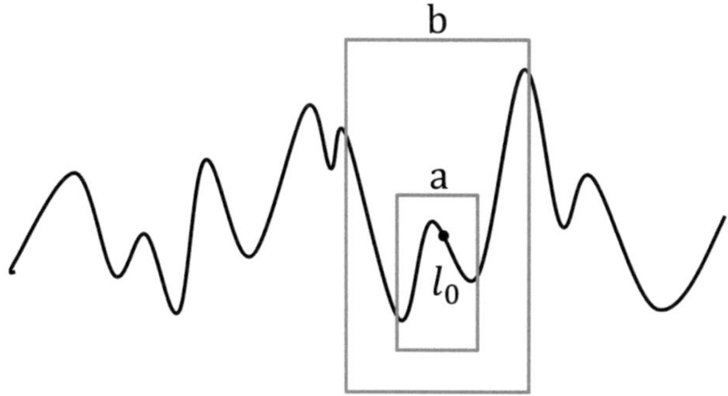


**Figure 5.3**   Scale-related morphological shapes in a rough curve.

**Definition 5.1 (Observation domain)** *Given a domain $\Gamma$ that does not contain any holes. $\Gamma$ is an observation domain of $l_0$, if and only if $l_0 \in \Gamma$ and $\Gamma \subset \Omega_g$. The size of $\Gamma$ represents the observation scale.*

**Definition 5.2 (Complete set of observation domains)** *A complete set of observation domains at the point $l_0$ is denoted as $U = \{\Gamma_k\}$ with $k = 1, 2, 3, \ldots, K$, which is composed of $\Gamma$ and all its translations that fulfill Definition 5.1.*

**Definition 5.3 (Peaks and valleys)** *Let $\Psi_k$ be the boundary of $\Gamma_k$, which is an observation domain of the point $l_0$. At an observed scale, the point $l_0$ is the member of a peak, if $\exists \Gamma_k \subset U$ for which $g(l_0)$ is larger than all values in $\{g(l_0), l_0 \in \Psi_k\}$. In contrary, the point $l_0$ is a member of a valley, if $g(l_0)$ is smaller than all values in $\{g(l_0), l_0 \in \Psi_k\}$.*

**Definition 5.4 ("Flat" and "rough" regions)** *"Flat" regions do not mean absolutely constant areas, but the points that are neither recognized as peaks nor as valleys in a range of observed scales. "Rough" regions refer to the points that are simultaneously recognized as peaks and valleys.*

**Definition 5.5 (Morphological stability)** *Given a range of observation scales, $[s_1, s_2]$, state a pixel is morphologically stable, if it is steadily* *recognized as a member of valleys, peaks or "flat" regions; otherwise it is located in one of the "rough" regions, which are morphologically unstable.*

Numerically, the visual process described above is equivalent to top- and bottom-hat transforms in a scale space. In Table 5.1 the topographical shapes are related to morphological feature curves. This idea etablishes the foundation of the algorithm for detecting graphite grains. In the next section a shape descriptor is developed for practical applications.

**Table 5.1**   Classification of topographical shapes.

| Shape types | $h(l,s), s \in [s_1, s_2]$ | $\check{h}(l,s), s \in [s_1, s_2]$ |
|:---:|:---:|:---:|
| Stable peak | Inconstant | Constant |
| Stable valley | Constant | Inconstant |
| "Rough" region | Inconstant | Inconstant |
| "Flat" region | Constant | Constant |

### 5.2.4 Application to shape description

It is intended to identify topographical shapes of image intensities by pairwise observation of $h(l,s)$ and $\check{h}(l,s)$. However, for a numerical analysis it is not convenient to consider feature pairs. In order to merge bright and dark features into a single measure, a signature for the morphological stability is developed in the following form:

$$\tilde{m}(l) = \frac{m(l) - \check{m}(l)}{m(l) + \check{m}(l)}, \tag{5.13}$$

where $m(l)$ and $\check{m}(l)$ are the steepness of feature curves, which are required to fulfill $m(l) + \check{m}(l) \gg 0$. There are two options to calculate $m(l)$ and $\check{m}(l)$. The first one is to employ the global steepness of peaks and valleys for this purpose:

$$m(l) = \frac{h(l,s_2) - h(l,s_1)}{s_2 - s_1}, \tag{5.14}$$

$$\check{m}(l) = \frac{\check{h}(l, s_2) - \check{h}(l, s_1)}{s_2 - s_1}.\tag{5.15}$$

Here, only the features at the lowest and the highest scales are considered for representing the feature variations. Secondly, details of feature curves can be taken into account:

$$m(l) = \frac{h(l, s_2) - h(l, s_1)}{s_2 - s_1 - s_0},\tag{5.16}$$

$$\check{m}(l) = \frac{\check{h}(l, s_2) - \check{h}(l, s_1)}{s_2 - s_1 - \check{s}_0},\tag{5.17}$$

where $s_0$ and $\check{s}_0$ are the length of scale intervals in which no bright and dark features are found. Now, the numerical criterion for classifying topographical shapes is presented:

- $\tilde{m}(l)$ is equal to $+1$ on stable peaks and $-1$ on stable valleys;

- $\tilde{m}(l) \in (-1, 1)$ indicates "rough" regions;

- $\tilde{m}(l)$ is not defined on "flat" zones where $m(l)$ and $\check{m}(l)$ is less than a threshold $T_m$.

Before $\tilde{m}(l)$ is computed, "flat" zones should be investigated separately. Morphological stability measures computed with the overall and detail-dependent steepness are denoted as $\tilde{m}_1(l)$ and $\tilde{m}_2(l)$, respectively.

## 5.3 Detection scheme

This section applies the developed shape descriptor to the graphite detection on laser-exposed surfaces. The usefulness of the method is explained by modeling the surface image in a photometric point of view. In the last step, the post-processing is designed to eliminate residual grooves from the segmented foreground objects. Graphite grains are to be detected at the remaining objects.

### 5.3.1 Feature extraction

In optical micrographs, graphite grains and deep grooves are shown as black regions. They are regarded as the foreground. The remaining image domain including metal shining and fine grooves comprises the background. The image formation process introduced in Section 3.1 inspires me to model LOM images of laser-exposed surfaces in a nonlinear fashion. Let $g(l)$ be a surface image, which can be formally written as

$$g(l) = \begin{cases} \zeta(l) & l \in \Omega_\zeta, \\ i(l)\tau(l) & l \in \Omega_\tau, \end{cases} \tag{5.18}$$

with $\Omega_g = \Omega_\zeta \cup \Omega_\tau, \Omega_\zeta \cap \Omega_\tau = \emptyset$. $\zeta(l)$ denotes the foreground. $i(l)$ decribes illumination non-uniformities. $\tau(l)$ represents an oscillatory background texture. In this model, image signals are nonlinearly connected, since physical surface structures occlude each other. Moreover, as an assumption the non-uniformity and oscillatory textures obey a multiplicative model. At low magnifications, optical micrographs of laser-exposed surfaces demonstrate the following photometric characteristics.

- Graphite grains have a non-metallic property, and residual grooves are deep structures. Thus, little light can be reflected from them into the imaging system. Foreground objects in $\zeta(l)$ are approximately invariant to the spatial illumination. Their intensities look like deep basins in image topography.

- Oscillatory textures are composed of metallic microfacets with random orientations. These microstructures lead to clustered intensity peaks and valleys in $\tau(l)$, which are sensitive to the design of the imaging system, e.g., the illumination mode and the specimen pose.

- In general, $i(l)$ varies much more slowly than other signal components. Hence, the texture signal multiplied with $i(l)$ still keeps the peak-valley structure.

Figure 5.4 illustrates these characteristics based on an image row. Surface components are manually marked in the image profile. In the following, it can be seen that morphological stability provides a scale-space evidence for graphite detection.

Then, top-hat and bottom-hat transforms are applied to the test image. The structuring elements are disk-shaped with radii increasing from 1 to

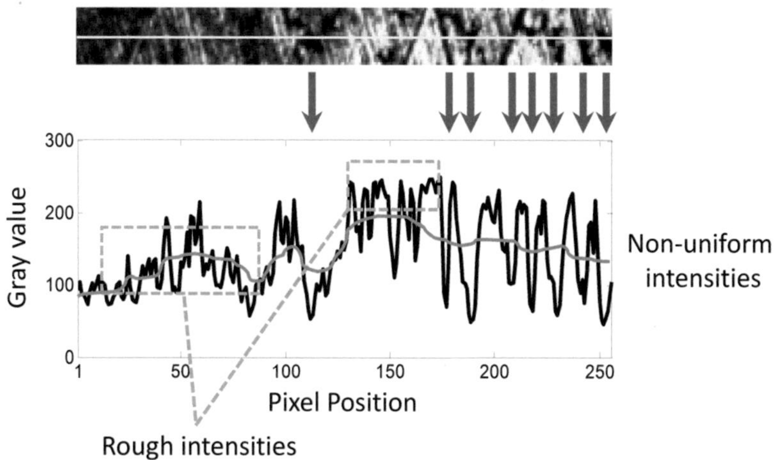


**Figure 5.4**  Profile of image intensities. Displayed intensities are located on the line marked in the gray image patch. The arrows indicate the foreground objects.

10 pixels. The choice of the scale range will be explained later. Figure 5.5 shows bright and dark feature curves of some pixel samples. For the foreground objects, the bright feature remains near zero in the top-hat scale space, whereas the dark feature increases in the bottom-hat scale space. In the textural background, local minima and maxima are leveled randomly. Therefore, stair-like feature curves appear in both scale spaces. By comparison, bright areas correspond to high peaks. Because of the duality of top-hat and bottom-hat transforms, the scale-space behaviors of shining areas are opposite to foreground objects.

In Fig. 5.5 bright and dark feature curves are shown in pairs. It can be found that they match the characteristics described in Table 5.1. During the foreground segmentation the emphasis is only put on intensity valleys. This allows me to rectify the signature of morphological stability by setting its positive and undefined positions to zero, that is,

$$\tilde{m}'(l) = \begin{cases} \tilde{m}(l) & \tilde{m}(l) < 0, \\ 0 & \tilde{m}(l) \geq 0, \\ 0 & m(l) < T_m, \check{m}(l) < T_m. \end{cases} \tag{5.19}$$

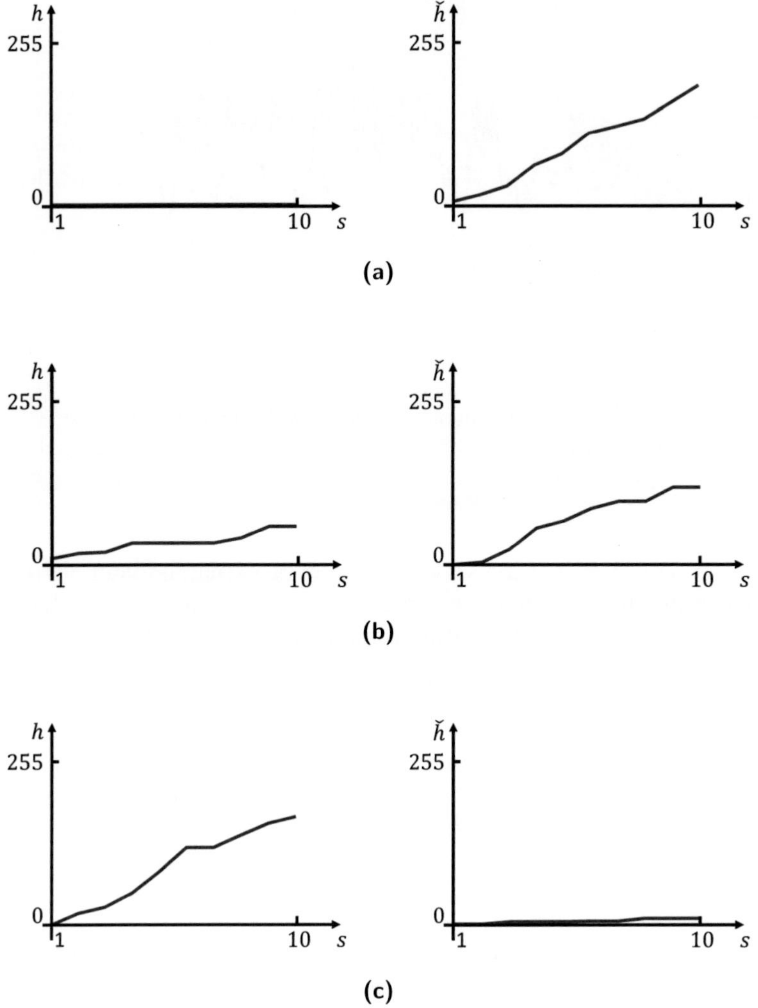

**Figure 5.5**  Dual observation of the image topography. (a) Valley. (b) Oscillatory texture. (c) Peak.

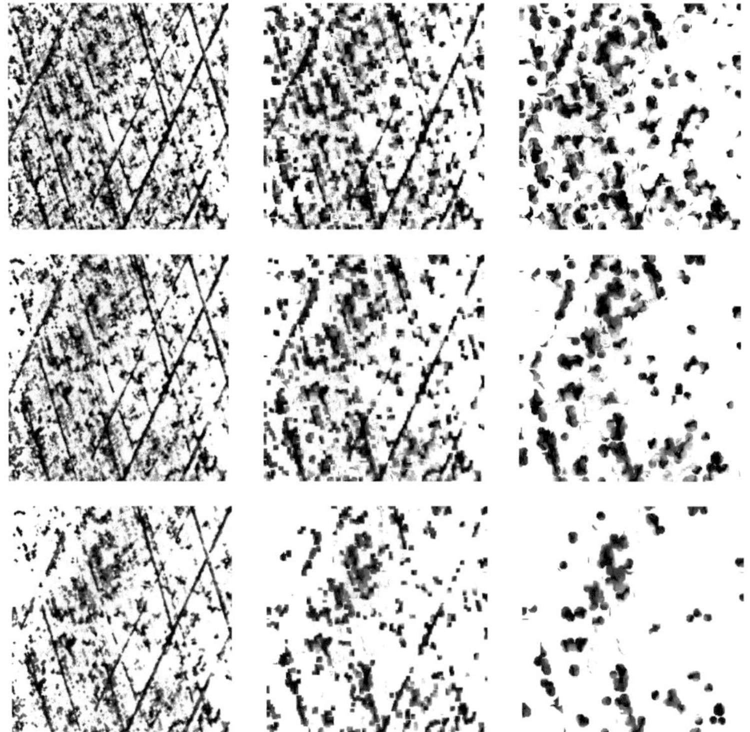

**Figure 5.6** Scale selection. Columns from left to right: $s_1 = 1, 3, 5$. Rows from top to bottom: $s_2 = 10, 15, 20$.

Accordingly, the rectified versions of $\tilde{m}_1(l)$ and $\tilde{m}_2(l)$ are notated as $\tilde{m}_1'(l)$ and $\tilde{m}_2'(l)$, respectively. For LOM images digitized with 256 gray levels, "flat regions" can be found by setting the threshold, $T_m$, to 5. This way, significant valleys cannot be lost by the rectification. Considering two options of the steepness measure, $\tilde{m}_1'(l)$ and $\tilde{m}_2'(l)$ are combined with a minimum operator:

$$\tilde{m}_g(l) = \min_l \left\{ \tilde{m}_1'(l), \tilde{m}_2'(l) \right\}. \tag{5.20}$$

Furthermore, the discussion comes to the influence of scale selection. Firstly, the size of $B(l)$ determines the step length for enlarging the structuring element. Given a disk-like structuring element $B(l)$ with a radius of $r_B$ pixels, the radius increment from $B_s(l)$ to $B_{s+1}(l)$ amounts $(2r_B - 1)$ pixels. In order to reveal as much topographical information as possible, the scale resolution is set to the minimum, i.e., $r_B = 1$. Secondly, the scale range with a lower limit, $s_1$, and an upper limit, $s_2$, is specified to sieve valleys of different scales. This property is shown by choosing $s_1 \in \{1, 3, 5\}$ and $s_2 \in \{10, 15, 20\}$. A matrix of feature maps is then obtained in Fig. 5.6. It can be seen that small particles are suppressed as $s_1$ becomes large, whereas large valleys in non-uniform intensities are extract by increasing $s_2$. Therefore, the scale range should be adjusted according to the magnification of LOM images. For the optical micrographs used in the experiments, $s_1 = 1$ and $s_2 = 10$ are considered to be moderate.

### 5.3.2 Segmentation

The feature image, $\tilde{m}_g(l)$, is segmented with Otsu's threshold. The foreground objects are located in the segmented regions owning the lowest value of $\tilde{m}_g(l)$. To identify graphite grains, the foreground should be further separated into grooves and graphite grains. Firstly, the morphological path opening introduced in Appendix B is used to find elongated objects that are not perfectly straight. The image grid is taken as an oriented graph. Paths defined in this graph are utilized as structuring elements. This step accomplishes a preliminary separation of grains and grooves. Since the path opening eliminates lines that contain paths larger than a given path length, short and thin lines are retained in the grain map. To deal with this problem, the binary maps separated by path opening are corrected by investigating the roundness of connected objects. Elongated objects are moved from the grain map to the groove map. Similarly, round particles are moved from the groove map to the grain map.

## 5.4 Experimental results

The detection scheme for graphite grains is validated in two stages. Firstly, the focus lies in the foreground segmentation. The approach is compared with local thresholding approaches and the alternating sequential filter (ASF). Appendix C.2 summarizes local adaptive thresholding methods. The ASF is defined as

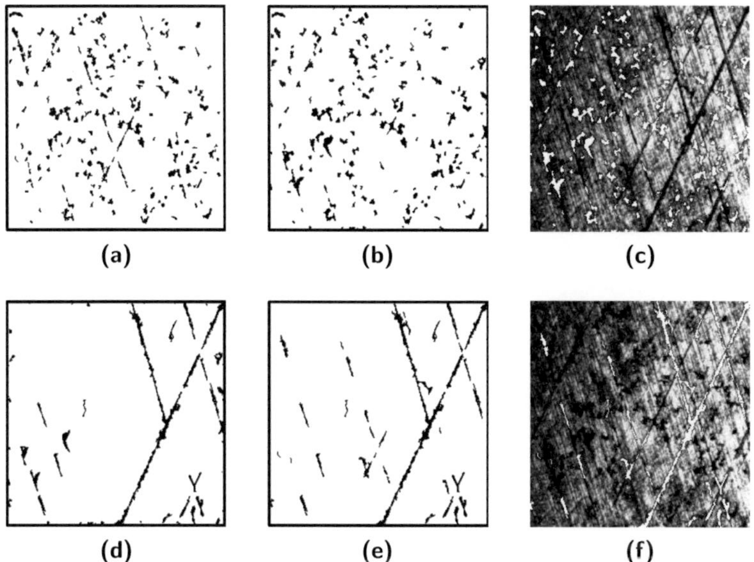

**Figure 5.7** Results of postprocessing. (a) and (d): Grains and grooves separated by path opening. (b) and (e): Corrected binary maps. (c) and (f): Grains and grooves marked in the original image.

$$a_s(l) = \left( \underbrace{g \circ B \bullet B \cdots B \circ B \bullet B}_{s \text{ times}} \right)(l). \tag{5.21}$$

In the experiment, $s$ is set to 20, and $B$ is a flat disk-like structuring element of radius 1. The ultimate detection results are verified in the second stage.

Figure 5.8 illustrates a series of foreground segmentations. Local adaptive thresholds and the ASF are resistant to non-uniform intensities, but are unable to tackle the problem induced by background grooves. The method is based on the fact that graphite particles, fine grooves and non-uniform intensities are present at different scales. This characteristic was exploited to reject background disturbances. Thus, both the lighting and textural disturbances have little impact on the segmentation result.

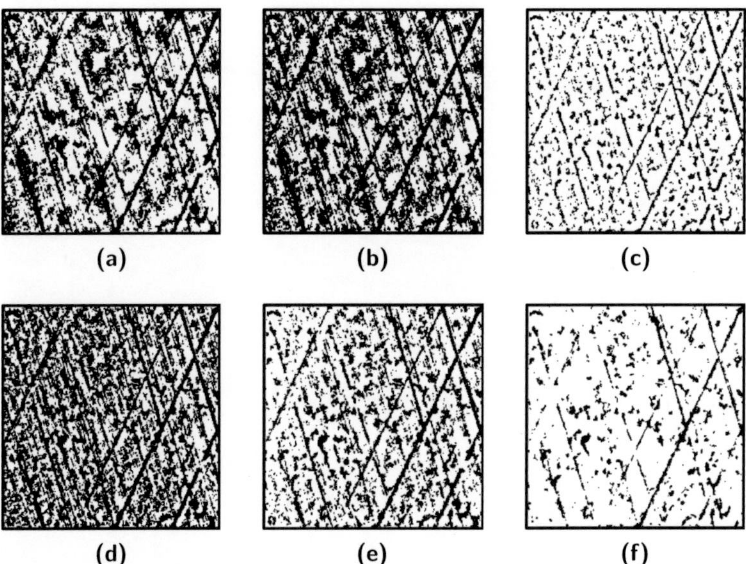

**Figure 5.8** Foreground segmentation. (a) ASF. (b) Bernsen. (c) Niblak. (d) Sauvola. (e) Local mean. (f) Method in this work.

Moreover, the correctness of graphite detection is verified by clipping graphite grains from the original image. Figure 5.7 illustrates that graphite grains adhered on coarse grooves are also eliminated, because they are also a part of the oil channel system. Additionally, some short groove fragments are still perceptible in the corrected grain map. Since their shapes are close to porous structures, they are considered as a part of the micro-chamber system. These experiments illustrate that the detection result can serve as a basis for an automated surface evaluation. More results for the validation can be found in Appendix A.2.

# 6 Surface evaluation

This chapter gains an insight into the quality parameters for the evaluation of 2D surface finishes. Table 6.1 indicates some quality measures for laser-exposed surfaces, which are mostly emphasized by engine producers. The quality measures for plateau-honed surfaces are incompletely summarized in Table 6.2. Some applications issues for these quality measures should be noticed. Firstly, some of these measures are related with the image contrast, the magnification or the field of view. They describe the visual impression of surface qualities. For this reason, an individual quality measure is comparable among surface samples only under specified imaging conditions. Furthermore, surface images should be acquired at different places on the cylinder wall during the non-100% inspection. The statistical analysis of quality measures for a group of surface samples is needed to evaluate the overall quality of cylinder bore surfaces. The mean and standard deviation of each quality measure are computed for the final evaluation. Based on the detection schemes presented in this thesis, another two quality parameters are developed. The usage of the new quality measures is as same as introduced before.

Table 6.1  Quality measures for laser-exposed cylinders.

| Surface components | Quality parameters |
| --- | --- |
| Grooves | Area of residual deep grooves. |
| Graphite grains | Size of grains, distribution uniformity, density of grains. |

Table 6.2 Quality measures for plateau-honed cylinders.

| Surface components | Quality parameters [2, 14, 59, 80] |
|---|---|
| Defects | Area of metal folds, overall degree of defectivity, area of bubbles and blowholes. |
| Groove texture | Honing angle, groove dispersion, balance of groove sets, amount of axial grooves, groove width, area of grooves and plateaus. area of groove interrupts. |

## 6.1 Defect severity

The proposed measure for defect severity describes the distribution of metal folds, which can be associated with the density and strength of defective edges. Unlike the overall degree of defectivity presented in [59] the damage of honing grooves is not considered in the defect severity. This is advantageous for the study of individual surface components. The measure of defect severity is developed in the following formula:

$$D = \frac{\sum_{\mathbf{x} \in \Omega_d} \lambda_1(\mathbf{x})}{A}. \tag{6.1}$$

The numerator is a sum of the strength of detective edges. $\Omega_d$ is the segmented defect region. The denominator, $A$, is the area of a rectangular image. Without loss of generality, all results presented in this section are measured in pixels. Figure 6.1 shows a series of surface samples visually sorted in the order from "seriously defective" to "slightly defective." The displayed images are of size $256 \times 256$. The detection results are also demonstrated in the same figure. In Fig. 6.2, the plot shows the values of

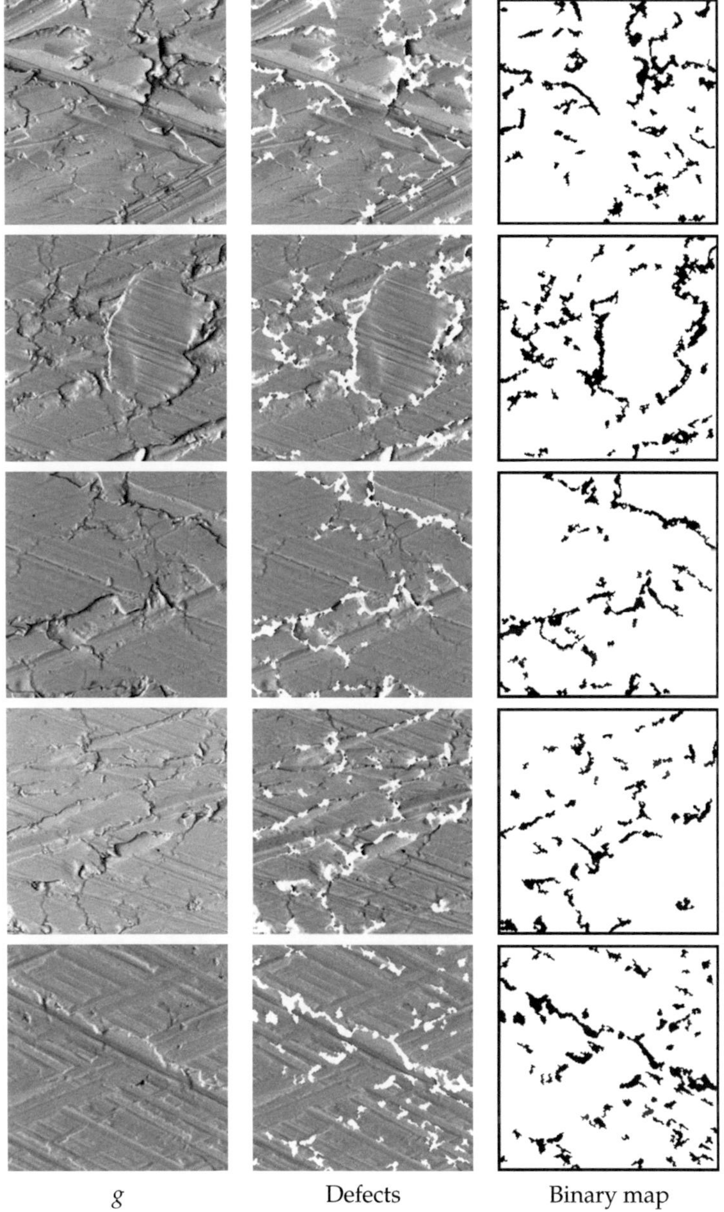

*g*　　　　　Defects　　　　　Binary map

**Figure 6.1**　Detection results of surface samples.

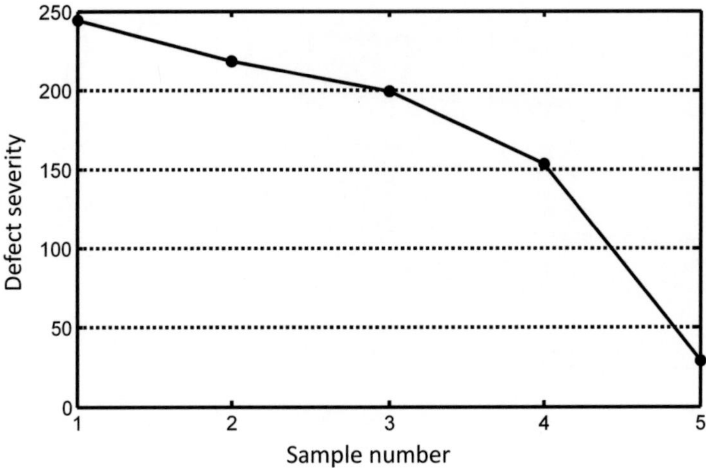

**Figure 6.2**  Surface evaluation in defect severity.

defect severity against the sample number. It can be seen that the result achieved by the numerical measure coincides with the visual impression.

## 6.2  Graphite distribution

In the segmented graphite map, isolated entities are taken as graphite grains. These entities are labeled through the connectivity analysis according to [39]. The geometric centers of the labeled regions yield a point pattern, which indicates grain locations. When graphite grains are clustered, lubricant oil will be unevenly distributed on the surface. Additionally, the size of graphite grains is also expected to be uniform. To characterize the graphite distribution, spatial statistics [19] is employed. Moran's autocorrelation index [42] (generally denoted as $I$) is used to describe the spatial distribution of graphite grains. In this work, the corresponding formula is

$$I = \frac{N_c \sum_p \sum_q w_{pq} \left( c_p - \bar{c} \right) \left( c_q - \bar{c} \right)}{\left( \sum_p \sum_q w_{pq} \right) \sum_p \left( c_p - \bar{c} \right)^2}.$$

(6.2)

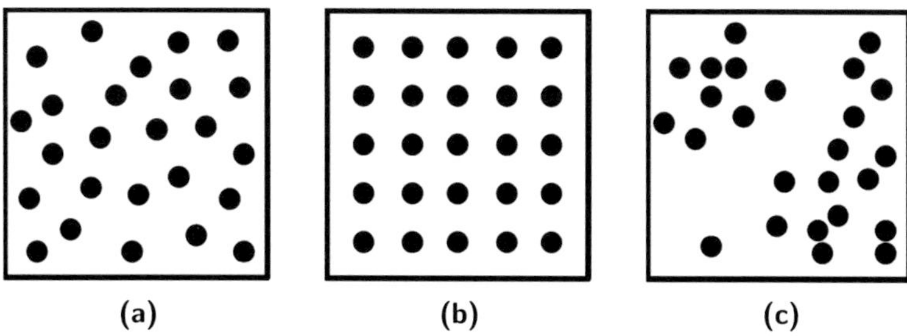

**Figure 6.3** Point patterns. (a) Random, $I = -1$. (b) Uniform, $I = -0.06$. (c) Cluster, $I = 0.78$. Random values from $[1, 100]$ are assigned to these points.

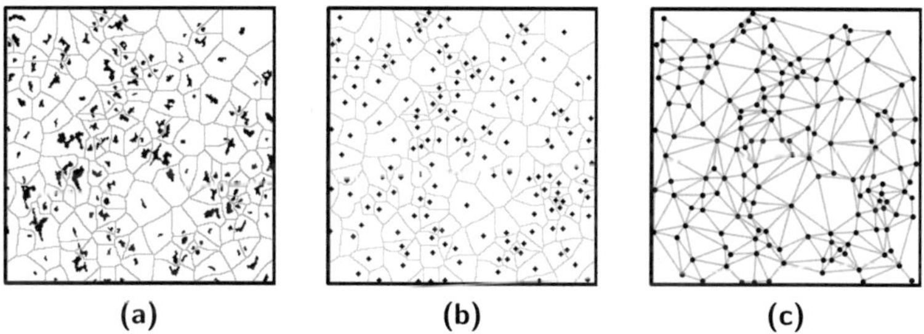

**Figure 6.4** Measures in Moran's index. $I = -0.83$. (a) Voronoi diagram. (b) Grain centers. (c) Delaunay triangles.

Here, $N_c$ is the number of detected graphite grains. $c_p$ and $c_q$ are two observations of a random variable $c$. In the present application, $c$ denotes the size of graphite grains, and $\bar{c}$ the mean grain size. $w_{pq}$ is the weight indicating the adjacency of observations at $p$ and $q$. As suggested in [24], $w_{pq}$ is chose to $1/d_{pq}$ for close neighbors, where $d_{pq}$ denotes the distance between neighboring grains. Otherwise, $w_{pq}$ is set to 0. Additionally, $w_{pp}$ is also set to 0 in the implementation. The adjacency relation of graphite grains is defined with a Voronoi diagram. Connecting neighboring graphite centers results in a map of Delaunay triangles, where the grain distance, $d_{pq}$, is measured with the side length of these triangles.

In the theory of spatial statistics, the spatial distribution of point patterns can be classified into three types, as shown in Fig. 6.3. As long as the variable values are inconstant at these points, the spatial relation works in Moran's index $I$. The expected value of Moran's index $I$ is equal to $-1/(N_c - 1)$. The uniform distribution has a high negative spatial autocorrelation, i.e., $I$ is significantly smaller than $-1/(N_c - 1)$ and near to $-1$. The random distribution does not show any spatial autocorrelation. In this case, $I$ is approximately equal to 0 if $N_c$ is very large. The cluster distribution indicates a high positive spatial autocorrelation. Thus, its $I$ value is close to 1. Figure 6.4 illustrates the process for implementing Moran's index $I$. The result shows that graphite grains contained in the test image are randomly distributed, which is desired for the laser-exposed honing.

# 7  Conclusions

This thesis has presented two image-based inspection approaches for the quality evaluation of cylinder bore surfaces. In the automotive industry, cylinder blocks are finished by honing, which can significantly improve the tribological performance of engines. The surface finish has a critical influence on engine lubrication and the wear of pistons. Therefore, cylinder bore surfaces are manufactured into specified textures which provide the desired surface functions. The work contributes to detect surface components and characterize these components for surface quality assessment.

Two types of cylinder bore surfaces have been investigated in this thesis. Firstly, defect detection was carried out on plateau-honed surfaces, where honing grooves serve as oil channels in the lubrication system. Secondly, the study concerned laser-exposed surfaces, where graphite grains form pores for reserving lubricant in running surfaces. To evaluate grooved and porous structures, different inspection strategies are proposed. On plateau-honed surfaces defective positions are located at cutting edges of tool marks, which are only perceptible in highly magnified images. Thus, the scanning electron microscopy was adopted for this purpose. The details of surface structures were explored in this part. On laser-exposed surfaces, graphite grains are shaped as scattered particles. To assess the graphite distribution, a low image magnification is needed to show an overview of the surface. Hence, image data used for this inspection task were captured by a light optical microscope. The work focused on image analysis algorithms which are suited for the aforementioned micrographs.

In the first algorithm developed in this thesis, metal folds were considered to be manufacturing failures on plateau-honed surfaces. These defects were actually metal burrs that could seriously wear pistons. Since metal folds were created in form of rough edges, defects on honed surfaces can be identified according to edge shapes. This needed an accurate estimation of edge positions and orientations on complex technical surfaces. Unfortunately, current orientation detectors were not good enough for the inspection task. To improve the performance of conventional gradient-based methods, edges and image noise were modeled in the squared-gradient field. Smoothing this vector field is fully identical with the eigenvalue analysis of the structure tensor. The problem was thus converted to constructing a structure-adaptive smoothing filter. The filter kernel was designed in a bilateral form. A pair of range filters was developed to achieve edge-preserving and orientation- selective properties. The one was attributed with gradient amplitudes, and the other relied on a set of features that were extracted by a Gabor-filter bank. Additionally, parameters for range filters were automatically selected with a novel method, which was based on a reasonable inference of image structures. Smoothing parameters were formulated according to profile shapes of sorted feature distances which were calculated in a local window. A novel structure tensor was constructed by extending the developed bilateral filter to the tensor field. It was then named as edge-aware structure tensor. The superior characteristics of this new orientation detector were validated with both synthetic and real images. Furthermore, a signature of the defect to identify rough edges was designed by combining orientation dispersion and defect saliency. The final segmentation of defective edges was accomplished with an automatic threshold. The visual and quantitative evaluation manifested that the proposed approach achieved the best results in comparison with the state-of-the-art techniques.

The second algorithm was designed for detecting graphite grains in optical micrographs. Laser-exposed surfaces are composed of spatially occluded surface components, which show different photometrical characteristics. Surface components invariant to the illuminant were taken as foreground objects, such as graphite grains and deep grooves, while the remaining plateau grooves constitute a background whose appearance varies with imaging conditions. These surface components can be described by features related to the image topography. Especially, foreground objects were always represented by deep intensity valleys in contrast to their neighborhood. This visual impression was numerically described with a metric for morphological stability. Top-hat and bottom-hat filters were useful for characterizing the image topography from above and below. This process tuned out to be the same as that in the visual recognition of graphite grains. It can be also noted that topographical shapes are scale-related. Hence, foreground objects were searched in a specific scale range. To detect peaks and valleys in the image topography, bright and dark features were extracted at each pixel location by hat transforms. Two morphological scale spaces obtained by relating these features with the size of the structuring element. These scale spaces could significantly illustrate topographical shapes of honed surfaces. Foreground objects were shown as stable valleys. Morphological stability was characterized by the steepness of peaks and valleys with respect to a same scale range. The boundaries of the detected peaks and valleys could be effectively determined based on the concept of morphological stability. Furthermore, the grain map was created by eliminating deep grooves from the binarized foreground map. An advanced morphological operator, path opening, was employed to remove elongated objects in binary images. The proposed method for detecting graphite grains was resistant to the disturbances induced by non-uniform illumination and background textures. The method was compared with local adaptive thresholding techniques and alternating sequential filters, which are commonly used for segmenting technical surfaces. The method was able to perform more stable segmentation of meaningful objects.

With the algorithm proposed above, a component-oriented evaluation of surface qualities was performed instead of a global assessment of surface patches. The reason for the component-oriented evaluation can be explained with an example. If plateau-honed surfaces are seriously damaged, expected cross-hatch textures are not successfully produced. In this case, textural features for globally evaluating honing grooves are less meaningful. To assess surfaces of a wide range of qualities, individual surface components have to be taken into account. In this thesis defect severity of plateau-honed surfaces was evaluated with the weighted spatial density of defects. The suitability of this measure was verified by sorting a series of damaged surface samples. Ranking the surface qualities based on the proposed quantitative measures led to the same results as the visual assessment. Moreover, the graphite distribution in laser-exposed surfaces was studied. Spatial auto-correlation of grain areas was taken as a quality measure to indicate the spatial uniformity of graphite grains. These quality parameters also verified the application potential of the image analysis algorithms.

In conclusion, the proposed algorithms can accurately and robustly separate images of cylinder bore surfaces into meaningful surface components. Even very challenging surfaces are able to be correctly evaluated in the segmentation results. At present, the inspection methods introduced in this work are still limited in laboratory applications. The extension to the inspection for mass production requires imaging systems which should be capable of generating high-quality images at large magnifications. Moreover, the EAST can be utilized for corner detection and diffusion filtering, which are typical applications of structure tensors. The measure of morphological stability provides a robust feature for objects with constant intensities. It will be useful for the segmentation tasks that rely on the size of objects rather than their intensity values.

# A  Results

## A.1  Defect inspection

The section presents additional experimental results that achieved by testing a series of surface samples. Feature images as well as the final defect map are calculated with the EAST-based detection scheme. Parameters are set to $\left(N_{\check{\xi}}, N, \kappa\right) = (18, 21, 0.2)$. These results show that the proposed algorithm can deal with a wide range of surface qualities.

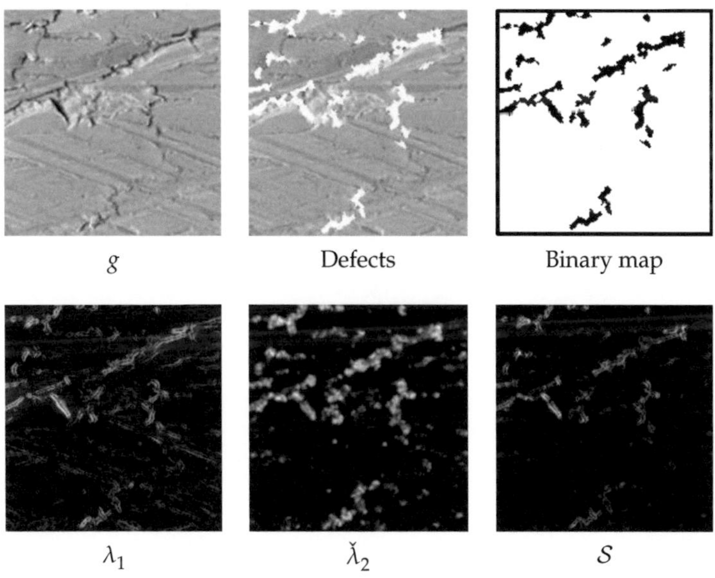

$g$      Defects     Binary map

$\lambda_1$     $\check{\lambda}_2$     $\mathcal{S}$

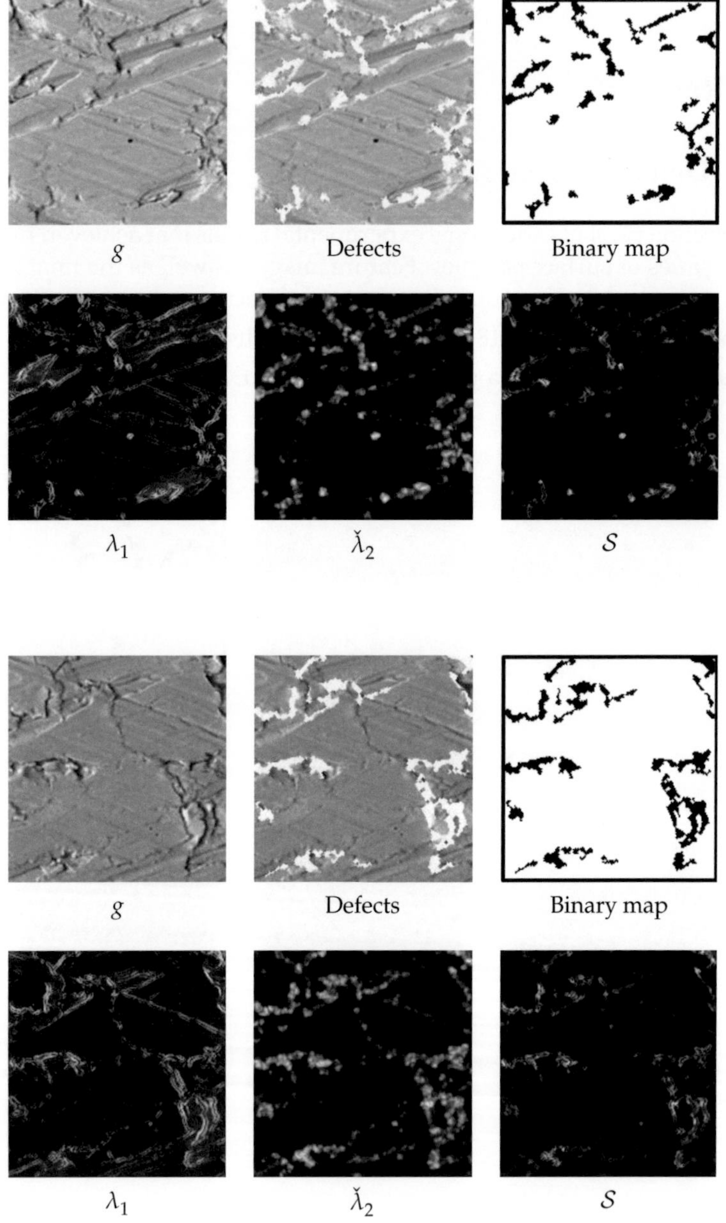

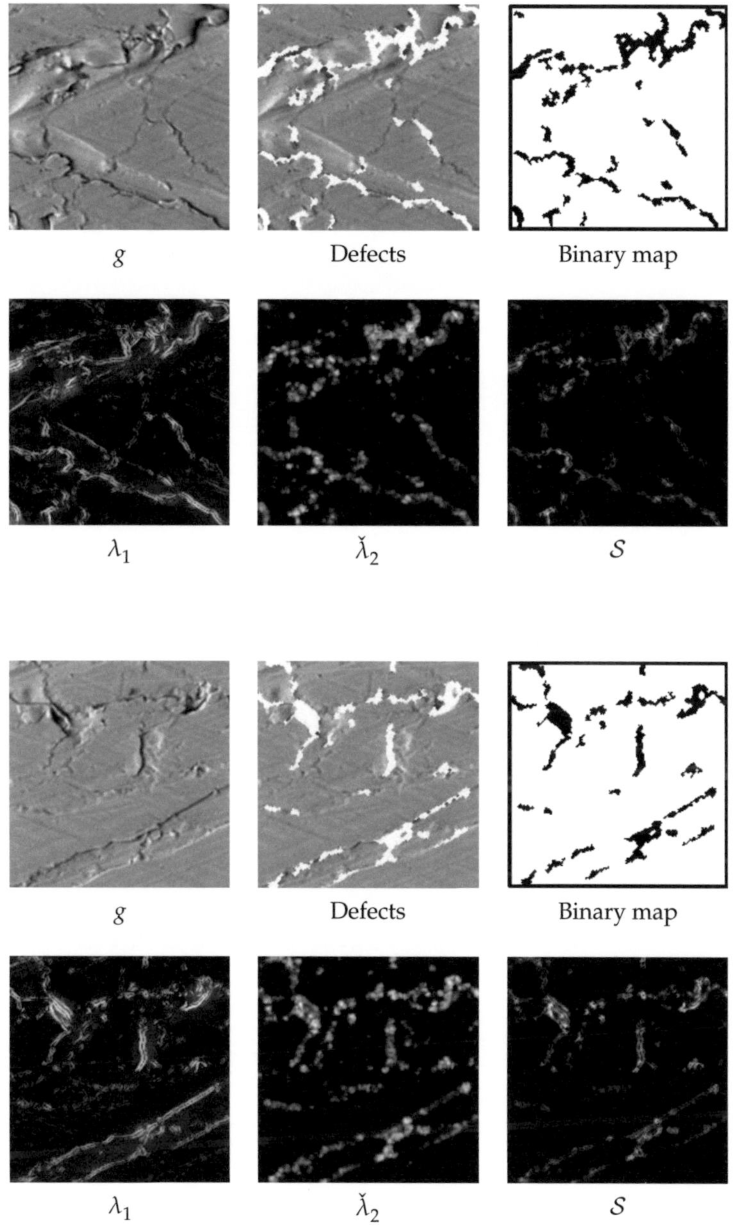

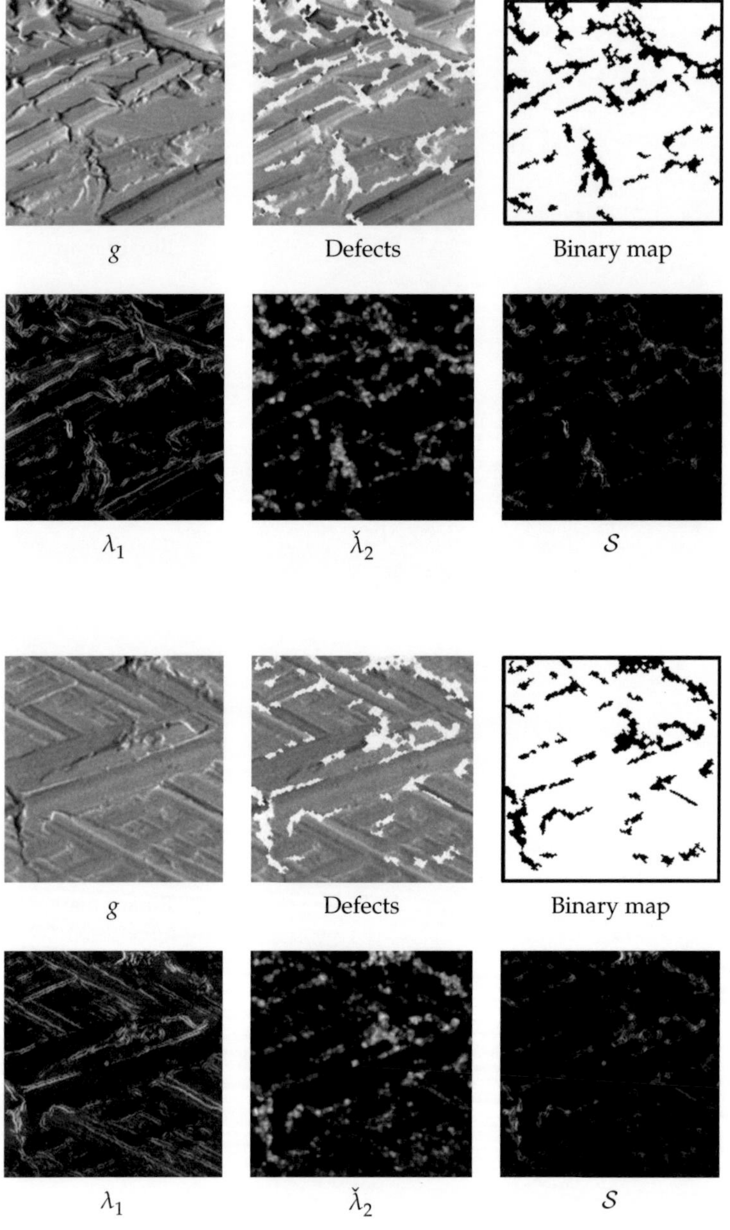
g
Defects
Binary map
λ₁
λ̌₂
S
g
Defects
Binary map
λ₁
λ̌₂
S

## A.2 Graphite detection

Several surface patches are examined in LOM images showing greatly varying illuminant conditions and background grooves. The detection scheme turns out to be very robust to these disturbances. The proposed method achieves good segmentation results in these experiments.

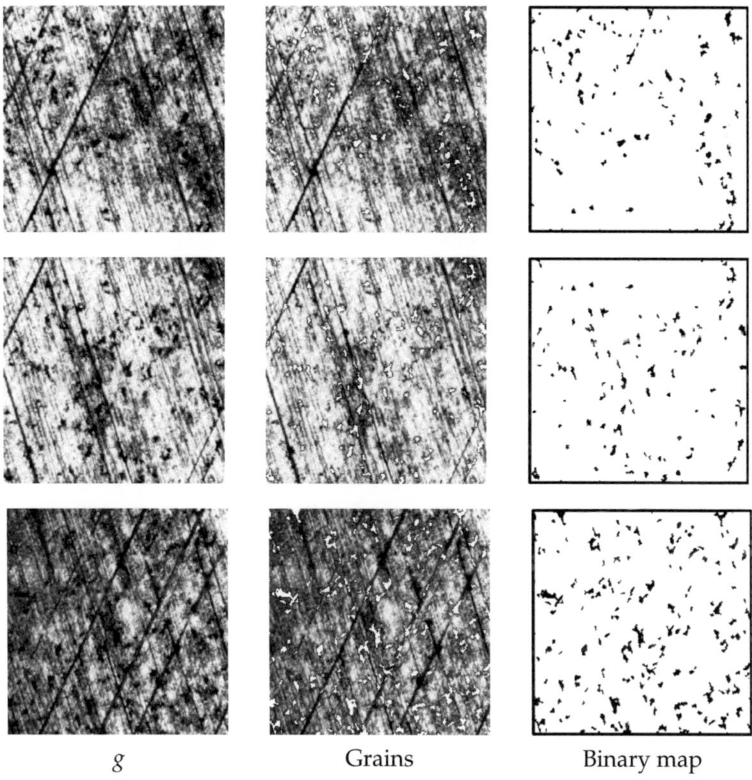

$g$        Grains        Binary map

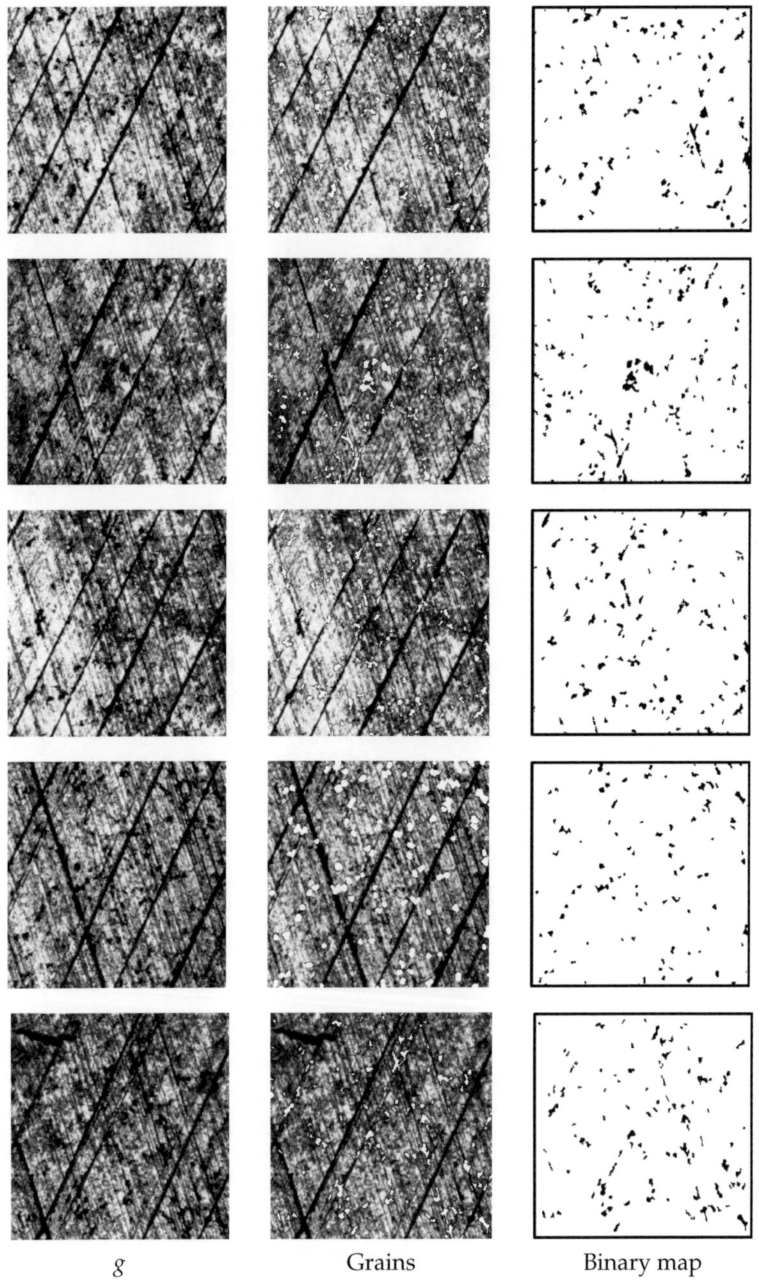

*g*      Grains      Binary map

# B Path openings and closings

Recently, advanced directional morphological operators, named as path openings and closings, were proposed in [27]. The conventional morphological method for detecting narrow and elongated objects used straight lines as structuring elements. Different from that, path openings and closings consider digital images as directed graphs. Structuring elements were specified with graph paths of a limited length. The benefits of doing so are obvious. On the one hand, line-like objects are not assumed to be perfectly straight. Some slightly curved lines are also detectable with path openings and closings. On the other hand, the thickness of line-like objects is not a critical issue by using path-based structuring elements. Both thin and thick lines can be found with a same structuring element. Formally, path openings and closings are defined as follows.

The principle is introduced with binary images. Let $\mathcal{G}$ be a directed graph endowed with an adjacent relation $z \mapsto z^*$. That means there is a graph edge from a node $z$ to another $z^*$. Here, $z$ is called a predecessor of $z^*$, and $z^*$ a successor of $z$. With these definitions, the dilation can be written as

$$\delta\left(\{z\}\right) = \left\{z \in \mathcal{G} | z \mapsto z^*\right\}. \tag{B.1}$$

In other words, the dilation of a subset $\chi \subseteq \mathcal{G}$ comprises all points which have a predecessor in $\chi$. Furthermore, the following notations are given. A $\delta$-path of length $L$ is defined as $b = (b_1, b_2, \ldots, b_L)$ with $b_{k+1} \in \delta\left(\{b_k\}\right)$ for $k = 1, 2, \ldots, L-1$. The elements of $b$ are contained in $\mathcal{P}(b) = \{b_1, b_2, \ldots, b_L\}$. The set of all $\delta$-paths of length $L$ is denoted as $\Pi_L$, in which the part contained in a subset $\chi$ of $\mathcal{G}$ is notated with $\Pi_L(\chi)$, i.e.,

$$\Pi_L(\chi) = \{b \in \Pi_L | \mathcal{P}(b) \subseteq \chi\}. \tag{B.2}$$

Consequently, the path opening can be defined as

$$\gamma_L(\chi) = \bigcup \{\mathcal{P}(b) | b \in \Pi_L(\chi)\}. \tag{B.3}$$

$\gamma_L(\chi)$ is the union of all paths of length $L$ contained in $\chi$. Conversely, the path closing is defined by exchanging the foreground and the background in a binary map.

Path openings and closings depend on the notion of graph connectivity. Figure B.1 demonstrates the adjacencies defined in [74], which are oriented in four different directions. Combination by supremum (for openings) and infimum (for closings) makes it possible to retain paths oriented in all possible directions just using these four adjacencies.

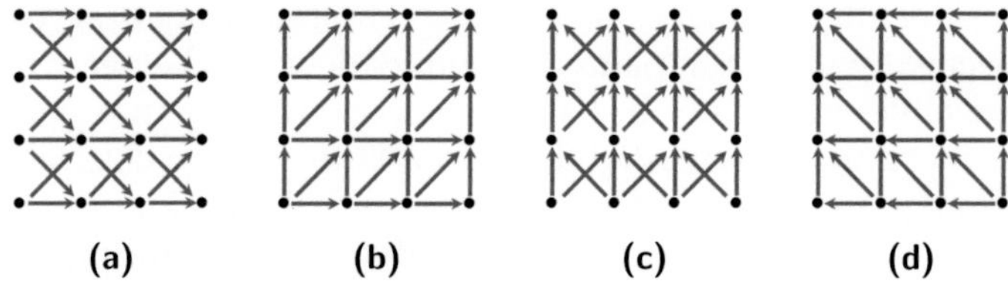

(a)        (b)        (c)        (d)

**Figure B.1**    Adjacencies in four directions. (a) East. (b) Northeast. (c) North. (d) Northwest.

# C Miscellaneous

## C.1 Roundness

The roundness, also named as the circularity, is the most important measure for graphite classification. As defined in [56], the roundness is expressed as

$$\text{roundness} = \frac{4c}{\pi D^2},$$ 

(C.1)

where $c$ is the area of the grain, and $D$ denotes the diameter of the smallest circle that can fully contain the grain. The graphical illustration in Fig. C.1 intuitively explains the meaning of roundness. As a shape parameter, the roundness is very useful for discriminating round and elongated objects.

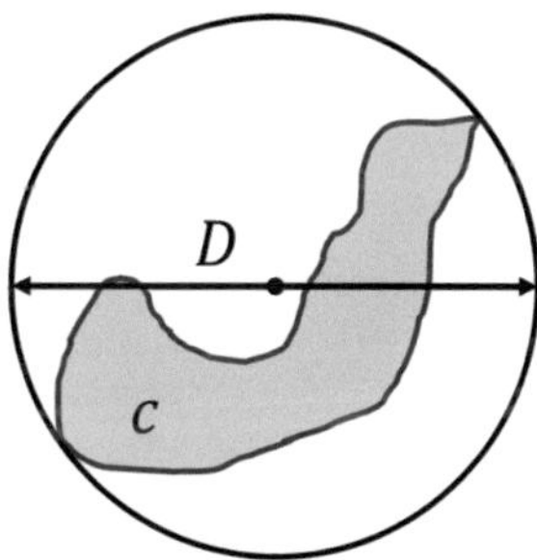


**Figure C.1**   Geometrical illustration of the roundness.

## C.2 Local adaptive thresholds

**Table C.1**  Local adaptive thresholds.

| Methods | Definition, $\mathbf{x} \in \Omega_{\mathbf{W}}$ |
| --- | --- |
| Bernsen | If $\max\{g(\mathbf{x})\} - \min\{g(\mathbf{x})\} > 15$, $T(\mathbf{x}) = 0.5Q$, where $Q = \max\{g(\mathbf{x})\} + \min\{g(\mathbf{x})\}$, else the background is segmented at $Q \geq 128$, and the foreground is segmented at $Q < 128$. The window size is $N = 31$. |
| Niblak | $T(\mathbf{x}) = \mu_{\mathbf{W}}(\mathbf{x}) + \beta\sigma_{\mathbf{W}}(\mathbf{x})$ with $\beta = -0.2$. The window size is $N = 15$. |
| Sauvola | $T(\mathbf{x}) = \mu_{\mathbf{W}}(\mathbf{x}) + 1 + \beta\left(\frac{\sigma_{\mathbf{W}}(\mathbf{x})}{128} - 1\right)$ with $\beta = -0.5$. The window size is $N = 15$. |
| Local mean | $T(\mathbf{x}) = \mu_{\mathbf{W}}(\mathbf{x})$. The window size is $N = 15$. |

Note: for images digitalized with 256 gray levels, $\mu_{\mathbf{W}}(\mathbf{x})$ is the local mean, and $\sigma_{\mathbf{W}}(\mathbf{x})$ is the local standard deviation. $\Omega_{\mathbf{W}}$ denotes the local window of size $N \times N$

# D Nomenclature

| Symbol | Description |
| --- | --- |
| | **Chapter 2: Honed surfaces** |
| $\alpha$ | Honing angle |
| $v_1$ | Horizontal speed of honing stones |
| $v_2$ | Vertical speed of honing stones |
| | **Chapter 3: Image acquisition** |
| $\phi$ | Incident angle of primary electron beam in SEM |
| | **Chapter 4: Defect detection** |
| $\mathbf{n}$ | Vector for optimization |
| $J$ | Objective function |
| $g$ | 1D or 2D gray-level image |
| $x$ | Spatial horizontal coordinate |
| $y$ | Spatial vertical coordinate |
| $\mathbf{x}$ | Vector spatial coordinate $(x, y)^\mathrm{T}$ |
| $N$ | Local window size |
| $\Omega_g$ | Support of $g$ |
| $\Omega_\mathbf{W}$ | Support of the local window |
| $\mathbf{T_C}$ | Classic structure tensor |
| $T_{11}, T_{12}, T_{21}, T_{22}$ | Elements of the structure tensor |
| $G_\mathbf{W}$ | Spatial filter kernel |
| $g_x$ | $x$-derivative of $g$ |
| $g_y$ | $y$-derivative of $g$ |
| $\nabla g$ | Gradient of $g$ |
| $|\cdot|$ | $L_1$-norm |

| Symbol | Description |
| --- | --- |
| $\lambda_1$ | The first eigenvalue of the structure tensor |
| $\lambda_2$ | The second eigenvalue of the structure tensor |
| det | Determinant of the structure tensor |
| tr | Trace of the structure tensor |
| $O$ | Angle of the first eigenvector corresponding to $\lambda_1$ |
| j | Imaginary unit |
| $r$ | Amplitude of the squared-gradient for $g\,(\mathbf{x})$ |
| $\varphi$ | Angle of the squared-gradient for $g\,(\mathbf{x})$ |
| $i$ | Non-uniform image intensity |
| $t$ | True texture signal in the image |
| $n$ | Noise signal in the image |
| $r_t$ | Amplitude of the squared-gradient for $t\,(\mathbf{x})$ |
| $\varphi_t$ | Angle of the squared-gradient for $t\,(\mathbf{x})$ |
| $r_n$ | Amplitude of the squared-gradient for $n\,(\mathbf{x})$ |
| $\varphi_n$ | Angle of the squared-gradient for $n\,(\mathbf{x})$ |
| $R_W$ | Amplitude of the mean squared-gradient for $g\,(\mathbf{x})$ |
| $\Phi_W$ | Angle of the mean squared-gradient for $g\,(\mathbf{x})$ |
| $R_t$ | Amplitude of the mean squared-gradient for $t\,(\mathbf{x})$ |
| $\Phi_t$ | Angle of the mean squared-gradient for $t\,(\mathbf{x})$ |
| $R_n$ | Amplitude of the mean squared-gradient for $n\,(\mathbf{x})$ |
| $\Phi_n$ | Angle of the mean squared-gradient for $n\,(\mathbf{x})$ |
| $\omega$ | Binary edge map |
| $\mathbf{u}$ | Mean squared-gradient |
| $\check{g}$ | Bilateral filter output |
| $G_\mathbf{R}$ | Range filter |
| $\mathbf{p}$ | Vector coordinate neighboring $\mathbf{x}$ |
| $C$ | Normalization factor in the bilateral filter |

| Symbol | Description |
| --- | --- |
| $g_0$ | Reference image for the cross bilateral filter |
| $f$ | Basic Gabor-filter in the spatial domain |
| $F$ | Basic Gabor-filter in the spatial frequency domain |
| $u, v$ | Coordinates of 2D Fourier domain |
| $\sigma_x, \sigma_y$ | Standard deviations in $f(x,y)$ |
| $\sigma_u, \sigma_v$ | Standard deviations in $F(u,v)$ |
| $\varsigma$ | Index for orientations in the Gabor-filter bank |
| $\eta$ | Index for scales in the Gabor-filter bank |
| $u_0$ | Central spatial frequency of $F(u,v)$ |
| $u_1$ | The lowest cantral frequency of scaled Gabor-filters |
| $u_{N_\eta}$ | The highest cantral frequency of scaled Gabor-filters |
| $\theta$ | The angle of Gabor filters |
| $\Delta\theta$ | Minimum angle interval of Gabor filters |
| $N_\varsigma$ | Orientation number of the Gabor-filter bank |
| $N_\eta$ | Scale number of the Gabor-filter bank |
| $x', y'$ | Scales and rotated $x$ and $y$ coordinats |
| $\rho$ | Scale factor |
| $f_{\varsigma,\eta}$ | Oriented and scaled Gabor-wavelet in the spatial domain |
| $\breve{g}_{\varsigma,\eta}$ | Output image filtered by $f_{\varsigma,\eta}$ |
| $\tilde{g}_\varsigma$ | Scale-invariant Gabor energy |
| $\tilde{\mathbf{g}}$ | Feature vector composed of all values of $\tilde{g}_\varsigma$ |
| $\|\cdot\|$ | $L_2$-norm |
| $G_{\sigma_1}$ | Range function relying on $|\nabla g|$ |
| $G_{\sigma_2}$ | Range function relying on $\|\tilde{\mathbf{g}}\|$ |
| $G_{\sigma_1\sigma_2}$ | Multiplication of $G_{\sigma_1}$ and $G_{\sigma_2}$ |

| Symbol | Description |
| --- | --- |
| $\sigma_1$ | Standard deviation for $G_{\sigma_1}$ |
| $\sigma_2$ | Standard deviation for $G_{\sigma_2}$ |
| $\mathbf{T}$ | Tensor at a pixel location |
| $\mathbf{T_E}$ | Edge-aware structure tensor |
| $\mathbf{T_B}$ | Bilateral structure tensor |
| $\mathbf{u_E}$ | Mean squared-gradient obtained from $\mathbf{T_E}$ |
| $k$ | Subscript for indexing variables |
| $d_1$ | Feature distance for computing $G_{\sigma_1}$ |
| $d_2$ | Feature distance for computing $G_{\sigma_2}$ |
| $\check{d}$ | Sorted array of feature distances |
| $H$ | Histogram of $\check{d}$ |
| $\mathbb{S}$ | Indices of $\check{d}$ labeled for structures similar to the window center |
| $\mathbb{A}$ | Indices of $\check{d}$ for ambiguous structures |
| $\mathbb{D}$ | Indices of $\check{d}$ for structures different from the window center |
| $l$ | Linear coordinate for array or matrix |
| $l_{\mathbb{A}}$ | Upper limit of $\mathbb{A}$ |
| $l_{\mathbb{S}}$ | Upper limit of $\mathbb{S}$ |
| $l_{\max}$ | Maxmum pixel number for sorting the feature distance in local windows |
| $l_1, l_2$ | Inflection points in the simplified profile of $\check{d}$ |
| $\kappa$ | Parameter in the estimation function for $\sigma_2$ |
| $\tilde{d}$ | Straight line connecting the begin and the end points of $\check{d}$ |
| $\mathcal{S}$ | Defect signature |
| $\check{\varphi}$ | Ground truth angle generated by smoothing $\varphi$ |
| $\mathcal{N}$ | Normalization function |

| Symbol | Description |
| --- | --- |
| $\check{\lambda}_2$ | Orientation dispersion calculated by the small eigenvalue |
| $A$ | Image area in pixel |
| $A_e$ | Area of segmented edge regions |
| $\Omega_e$ | Support of segmented edge regions |
| $\sigma_n$ | Standard deviation of image noise |
| **Chapter 5: Graphite detection** | |
| $l_x$ | Linear coordinate in the neighborhood of $l$ |
| $l_0$ | Observed signal location |
| $\mathbb{R}$ | Real number domain |
| $y_o$ | Filtered signal by morphological opening |
| $y_c$ | Filtered signal by morphological closing |
| $h$ | Output of the top-hat transform |
| $\check{h}$ | Output of the bottom-hat transform |
| $s$ | Scale index |
| $s_1$ | Lower limit of the scale range |
| $s_2$ | Upper limit of the scale range |
| $B$ | Structuring element |
| $\Delta h$ | Bright feature |
| $\Delta \check{h}$ | Dark feature |
| $\Gamma$ | Observation domain |
| $U$ | Complete set of the observation domain |
| $K$ | Size of U |
| $\Psi_k$ | Boundary of $\Gamma$ |
| $m$ | Steepness of peaks |
| $\check{m}$ | Steepness of valleys |
| $s_0$ | Total length of scale intervals where no bright feature is output |

| Symbol | Description |
| --- | --- |
| $\check{s}_0$ | Total length of scale intervals where no dark feature is output |
| $\tilde{m}$ | Morphological stability |
| $\tilde{m}_1, \tilde{m}_2$ | Optional expressions of $\tilde{m}$ |
| $\zeta$ | Foreground in $g$ |
| $\tau$ | Background texture in $g$ |
| $\Omega_\zeta$ | Support of $\zeta$ |
| $\Omega_\tau$ | Support of $\tau$ |
| $\tilde{m}'$ | Rectified $\tilde{m}$ |
| $\tilde{m}'_1, \tilde{m}'_2$ | Optional expressions of $\tilde{m}'$ |
| $\tilde{m}_g$ | Modified morphological stability |
| $r_B$ | Radius of $B$ |
| $a_s$ | Output of the alternating sequential filter at the $s$-th scale |
| $T$ | Local adaptive threshold |
| $\rho$ | Coefficient adjusting $T$ |
| $Q$ | Sum of local minimum and local maximum of $g$ |
| $\mu_W$ | Local mean of image intensities |
| $\sigma_W$ | Local standard deviation of image intensities |
| **Chapter 6: Surface evaluation** | |
| $\mathcal{D}$ | Defect severity |
| $\Omega_d$ | Support of segmented defective regions |
| $I$ | Moran's autocorrelation index |
| $p, q$ | Indice of grains |
| $w_{pq}$ | Weight in Moran's $I$ |
| $c$ | Grain area |
| $c_p, c_q$ | A pair of observations of |
| $\bar{c}$ | Mean grain area |

| Symbol | Description |
| --- | --- |
| $N_c$ | Number of graphite grains |
| $d_{pq}$ | Distance between a pair of graphite grains |
| | **Appendix B: Path openings and closings** |
| $z$ | Graph node as the predecessor |
| $z^*$ | Graph node as the successor |
| $\mathcal{G}$ | Oriented graph defined on the image grid |
| $\chi$ | A subset of $\mathcal{G}$ |
| $\delta$ | Node set containing all successors |
| $\boldsymbol{b}$ | Node vector denoting a path |
| $b_k$ | Node in the path $\boldsymbol{b}$ |
| $L$ | Path length |
| $\Pi_L$ | Path set containing all paths of length $L$ |
| $\gamma$ | Output of the path opening |
| $\mathcal{P}$ | Node set containing all nodes in the path $\boldsymbol{b}$ |
| | **Appendix C: Miscellaneous** |
| $D$ | Diameter of the smallest circle that fully contains a grain |
| $\beta$ | Coefficient for scaling local adaptive thresholds |

# Bibliography


[1]   **Ayres, F. J. and Rangayyan, R. M.** *Performance analysis of oriented feature detectors.* In: *Proceedings of the XVIII Brazilian Symposium on Computer Graphics and Image Processing.* Natal, RN, Brazil, 2005.

[2]   **Beyerer, J., Krahe, D., and Puente León, F.** *Characterization of cylinder bores.* In: *Metrology and Properties of Engineering Surfaces* (2001), pp. 243–282.

[3]   **Beyerer, J. and Puente León, F.** *Detection of defects in groove textures of honed surfaces.* In: *International Journal of Machine Tools and Manufacture* 37 (1997), pp. 371–389.

[4]   **Beyerer, J. and Puente León, F.** *Suppression of inhomogeneities in images of textured surfaces.* In: *Optical Engineering* 36.3 (1997), pp. 85–93.

[5]   **Beyerer, J. and Puente León, F.** *Adaptive separation of random lines.* In: *Optical Engineering* 37.10 (1998), pp. 2733–2741.

[6]   **Bien, H., Parikh, P., and Entcheva, E.** *Lenses and effective spatial resolution in macroscopic optical mapping.* In: *Physics in Medicine and Biology* 52.4 (2007), pp. 941–960.

[7]   **Boomgaard, B. T. R. van den.** *Adaptive structure tensors and their applications.* Tech. rep. Saarbrücken, Germany: Universität des Saarlandes, 2005.

[8]   **Brox, T.** *Nonlinear structure tensors.* Tech. rep. Saarbrücken, Germany: Universität des Saarlandes, 2004.

[9]   **Chen, J., Sato, Y., and Tamura, S.** *Orientation space filtering for multiple orientation line segmentation.* In: *IEEE Transactions on Pattern Analysis and Machine Intellegence* 22.5 (2000), pp. 417–429.

[10]   **Coulon, O., Alexander, D. C., and Arridge, S.** *Diffusion tensor magnetic resonance image regularization.* In: *Medical Image Analysis* 8 (2004), pp. 47–67.

[11]   **Davidson, M. W. and Mortimer, A.** *Optical microscopy.* 1999. URL: http://micro.magnet.fsu.edu/primer/opticalmicroscopy.html.

[12] **Decenciére, E. and Jeulin, D.** *Morphological decomposition of the surface topography of an internal combustion engine cylinder to characterize wear.* In: *Wear* 249.3 (2001), pp. 482–488.

[13] **Dienwiebel, M. and Scherge, M.** *Neue Erkenntnisse zur Tribologie von übereutektischen AlSi–Zylinderlaufflächen.* In: *Motortechnische Zeitschrift (MTZ)* 3 (2007), pp. 186–189.

[14] **Dimkovskia, Z.** *Characterisation of worn cylinder liner surfaces by segmentation of honing and wear scratches.* In: *Wear* 217.3–4 (2011), pp. 548–552.

[15] **Dore, V., Moghaddam, R. F., and Cheriett, M.** *Non–local adaptive structure tensors application to anisotropic diffusion and shock filtering.* In: *Image and Vision Computing* 29 (2011), pp. 730–743.

[16] **Falconer, K.** *Mathematical foundations and applications.* John Wiley & Sons, Ltd., 2003.

[17] **Felsberg, M., Kalkan, S., and Krüger, N.** *Continuous dimensionality characterization of image structures.* In: *Image and Vision Computing* 27.6 (2009), pp. 628–636.

[18] **Flores, G., Abeln, T., and Klink, U.** *Funktionsgerechte Endbearbeitung von Zylinderbohrungen aus Gusseisen.* In: *Motortechnische Zeitschrift (MTZ)* 3 (2007), pp. 181–185.

[19] **Fortin, M.-J., Dale, M., and Hoef, J. ver.** *Spatial analysis in ecology.* In: *Encyclopedia of Environmetrics* 4 (2002), pp. 2051–2058.

[20] **Freeman, W. T. and Andelson, E. H.** *The design and use of steerable filters.* In: *IEEE Transaction on Pattern Analysis and Machine Intelligence* 13.9 (1991), pp. 891–906.

[21] **Fritz, A. H. and Schulze, G.** *Fertigungstechnik.* 7. überarb. Aufl. Berlin, Heidelberg: Springer, 2006.

[22] **Geels, K.** *Light Microscopy, Image Analysis and Hardness testing.* In: ASTM International, 2006. Chap. Metallographic and materialographic specimen preparation.

[23] **Ginkel, M. van.** *Image analysis using orientation space based on steerable filters.* PhD thesis. Netherlands: University of Delft, 2002.

[24] **Gittleman, J. L. and Kot, M.** *Adaptation: statistics and a null model for estimating phylogenetic effects.* In: *Systematic Zoology* 39 (1990), pp. 227–241.

[25] **Han, J. and Ma, K.-K.** *Rotation–invariant and scale–invariant Gabor features for texture image retrieval.* In: *Image and Vision Computing* 25.1 (2007), pp. 1474–1481.

[26] **He, K., Sun, J., and Tang, X.** *Rotation–invariant and scale–invariant Gabor features for texture image retrieval.* In: *11th European Conference on Computer Vision.* Crete, Greece, 2010.

[27] **Hendriks, C. L.** *Constrained and dimensionality–independent path openings.* In: *IEEE Transactions on Image Processing* 19.6 (2010), pp. 1587–1595.

[28] *HOMMEL-ETAMIC IPS100 Innenprüfsensor.* Jenoptik AG. URL: http://www.jenoptik.com.

[29] **Jähne, B.** *Digtale Bildverarbeitung.* Heidelberg: Springer: Springer, 2005.

[30] **Jiang, H., Tan, Y., and Lei, J. F.** *Auto-analysis system for graphite morphology of grey cast iron.* In: *Journal of Automated Methods & Management in Chemistry* 25.4 (2003), pp. 87–92.

[31] **Jiang, X.** *On orientation and anisotropy estimation for online fingerprint authentication.* In: *IEEE Transactions on Signal Processing* 53.10 (2005), pp. 4038–4049.

[32] **Jiang, X.** *Extracting image orientation feature by using integration operator.* In: *Pattern Recognition* 40 (2007), pp. 705–717.

[33] **Kamarainen, J.-K., Kyrki, V., and Kälviäinen, H.** *Invariance properties of Gabor filter–based features—overview and applications.* In: *IEEE Transactions on Image Processing* 15.5 (2006), pp. 1088–1099.

[34] **Kamarainen, J.-K., Kyrki, V., and Kälviäinen, H.** *Local and global Gabor features for object recognition.* In: *Pattern Recognition and Image Analysis* 17.1 (2007), pp. 93–105.

[35] *Laserstrukturieren–Verbessern der tribologischen Eigenschaften von Oberflächen.* Gehring GmbH. 2004.

[36] **Lenhof, U. and Zwein, F.** *Messtechnik für Motorenzylinder.* In: *Motortechnische Zeitschrift (MTZ)* 5 (2002), pp. 360–369.

[37] **Li, X.** *Directional Gaussian derivative filter based palmprint authentication.* In: *International Conference on Computational Intelligence and Security.* Suzhou, China, 2008.

[38] **Manjunath, B. and Ma, W.** *Textural feature for browsing and retrieval.* In: *IEEE Transactions on Pattern Analysis and Machine Intelligence* 18.8 (1996), pp. 837–842.

[39] **Maragos, P.** *Image and Video Processing Handbook.* In: 2nd ed. Elsevier Academic Press, 2005. Chap. Morphological filtering for image enhancement and feature Detection, pp. 135–156.

[40] **Mester, R.** *Orientation estimation: conventional techniques and a new non–differental approach.* In: *Proceeding 10th European Signal Processing Conference.* 2000.

[41] **Michelet, F.** *Local multiple orientation estimation: isotropic and recursive oriented network.* In: *17th International Conference on Pattern Recognition.* Cambridge, UK, 2004.

[42] **Moran, P. A. P.** *Notes on continuous stochastic phenomena.* In: *Biometrika* 37.1 (1950), pp. 17–23.

[43] **Movellan, J.** *Tutorial on Gabor filters.* 2008. URL: http://mplab.ucsd.edu/tutorials/pdfs/gabor.pdf.

[44] **Mukhopadhyay, S. and Chanda, B.** *A multiscale morphological approach to local contrast enhancement.* In: *Signal Processing* 80 (2000), pp. 685–696.

[45] *MVplus cylinder inspector.* Wolf Systeme AG. URL: http://www.wolfsysteme.de.

[46] **Nath, S. K. and Palaniappan, K.** *Adaptive robust structure tensors for orientation estimation and image segmentation.* In: *Proceedings of the First international conference on Advances in Visual Computing.* Lake Tahoe, NV, USA, 2005.

[47] **Nikolaidis, N. and Pitas, I.** *Nonlinear processing and analysis of angular signals.* In: *IEEE Transactions on Signal Processing* 46.12 (1998), pp. 3181–3194.

[48] *Oberflächenscanner.* Breitmeier Messtechnik GmbH. URL: http://www.breitmeier.de.

[49] **Osher, J.** *Bildanalytische Charakteriesierung von Graphit im Grauguss und Klassifikation der Lamellenanordnung.* In: *Practical Metallography* 40.9 (2003), pp. 454–473.

[50] **Otsu, N.** *A threshold selection method from gray–level histograms.* In: *IEEE Transaction on Systems, Man, and Cybernetics* 9.1 (1979), pp. 62–66.

[51]  **Paris, S.** *Bilateral filtering: theory and applications.* In: *Computer Graphics and Vision* 4.1 (2009), pp. 1–73.

[52]  **Pehnelt, S., Osten, W., and Seewig, J.** *Vergleichende Untersuchung optischer Oberflächenmessgeräte mit einem Chirp-Kalibriernormal.* In: *Technisches Messen* 78.10 (2007), pp. 405–421.

[53]  **Pei, S. C., Lai, C. L., and Shih, F. Y.** *An efficient class of alternating sequential filters in morphology.* In: *Graphical Models and Image Processing* 59.2 (1997), pp. 109–116.

[54]  **Petschnigg, G.** *Digital photography with flash and no–flash image pairs.* In: *ACM Trans. Graph* 23.3 (2004), pp. 664–672.

[55]  **Pouliquen, F. L.** *A new adaptive framework for unbiased orientation estimation in textured images.* In: *Pattern Recognition* 38 (2005), pp. 2032–2046.

[56]  **Prakash, P., Mytri, V. D., and Hiremath, P. S.** *Classification of cast iron based on graphite grain morphology using simple shape descriptors.* In: *International Journal of Engineering and Technology* 2.4 (2009), pp. 37–42.

[57]  **Prakash, P., Mytri, V. D., and Hiremath, P. S.** *Digital microstructure analysis system for testing and quantifying the ductile cast iron.* In: *International Journal of Computer Applications* 19.3 (2011), pp. 22–27.

[58]  **Puente León, F.** *Evaluation of cylinder bores.* In: *Annals of CIRP* 51.1 (2002), pp. 503–506.

[59]  **Puente León, F.** *An objective measure of the quality of honed surfaces.* In: *Optical Measurement Systems for Industrial Inspection IV* 5856 (2005), pp. 287–295.

[60]  **Puente León, F. and Beyerer, J.** *Oberflächencharakterisierung durch morphologische Filterung.* In: *Technisches Messen* 72 (12), pp. 663–670.

[61]  **Radzikowska, J. M.** *Metallography and Microstructures.* In: vol. 9. ASM Handbook. ASM International, 2004. Chap. Metallography and microstructures of cast iron, pp. 565–587.

[62]  **Reimer, L. and Pfefferkorn, G.** *Raster–Elektronenmikroskopie.* Berlin: Springer, 1999.

[63]  **Ribeiro, E. and Shah, M.** *Computer vision for nanoscale imaging.* In: *Machine Vision and Applications* 17 (2006), pp. 147–162.

[64] **Robota, A. and Zwein, F.** *Einfluss der Zylindertopographie auf den Ölverbrauch und die Partikelemissionen eines DI–Dieselmotors.* In: *Motortechnische Zeitschrift (MTZ)* 4 (1999), p. 246.

[65] **Roos, E. and Maile, K.** *Werkstoffkunde für Ingenieure: Grundlagen, Anwendung, Prüfung.* Berlin, Heidelberg: Springer, 2002.

[66] **Scheib, H.** *Untersuchung des Zusammenhangs zwischen Erstarrung, Abkühlung, Gefüge und mechanischen Eigenschaften von dünnwandigen GJV Gussbauteilen.* PhD thesis. Otto-von-Guericke-Universität Magdeburg, 2007.

[67] **Schmid, J.** *Optimiertes Honverfahren für Gusseisen-Laufflächen.* In: *VDI-Tagungsband Zylinderlaufbahn, Kolben, Pleue.* 2006.

[68] **Schmid, J.** *Reibungsoptimierung von Zylinderlaufflächen aus Sicht der Fertigungstechnik.* In: *Motortechnische Zeitschrift (MTZ)* 6 (2010), pp. 408–413.

[69] **Seck, E. and Strobel, J.** *Diamant–Fluidstrahl: Ein neues Verfahren zur Bearbeitung der Zylinderlaufbahnen von Kurbelgehäusen aus Grauguss.* In: *Motortechnische Zeitschrift (MTZ)* 1 (2001), pp. 184–189.

[70] **Sezgin, M. and Sankur, B.** *Survey over image thresholding techniques and quantitative performance evaluation.* In: *Journal of Electronic Imaging* 13.1 (2004), pp. 146–165.

[71] **Swets, J. A.** *Signal detection theory and ROC analysis in psychology and diagnostics: Collected papers.* Lawrence Erlbaum Associates, 1995.

[72] **Tomasi, C. and Manduchi, R.** *Bilateral filtering for gray and color images.* In: *Procedings of the Sixth International Conference on Computer Vision.* Bombay, India, 1998.

[73] **Tsai, D. M.** *Optimal Gabor filter design for texture segmentation using stochastic optimization.* In: *Image and Vision Computing* 19 (2001), pp. 299–316.

[74] **Valeroa, S.** *Advanced directional mathematical morphology for the detection of the road network in very high resolution remote sensing images.* In: *Pattern Recognition Letters* 31.10 (2010), pp. 1120–1127.

[75] **Voort, G. F. V.** *Failure analysis and prevention.* In: vol. 11. ASM Handbook. ASM International, 2002. Chap. Metallographic techniques in failure analysis, pp. 498–515.

[76] **Vovk, U., Pernus, F., and Likar, B.** *A review of methods for correction of intensity inhomogeneity in MRI.* In: *IEEE Transactions on Medical Imaging* 26.3 (2007), pp. 405–421.

[77] **Wang, Y., Hu, J., and Han, F.** *Enhanced gradient–based algorithm for the estimation of fingerprint orientation fields.* In: *Applied Mathematics and Computation* 185.1 (2007), pp. 823–833.

[78] **Weidner, A., Seewig, J., and Reithmeier, E.** *3D roughness evaluation of cylinder liner surfaces based on structure-oriented parameters.* In: *Measurement Science and Technology* 17.3 (2006), pp. 477–482.

[79] **Wong, A.** *Adaptive bilateral filtering of image signals using local phase characteristics.* In: *Signal Processing* 88.1 (2008), pp. 1615–1619.

[80] **Xin, B.** *Automatische Auswertung und Charakterisierung dreidimensionaler Messdaten technischer Oberflächen mit Riefentexturen,* PhD thesis. Karlsruher Institut für Technologie, 2008.

[81] **Zhang, L., Zhang, L., and Zhang, D.** *A multi–scale bilateral structure tensor based corner detector.* In: *9th Asian Conference on Computer Vision.* Xi'an, China, 2009.

[82] **Zhang, M. and Gunturk, B. K.** *Multiresolution bilateral filtering for image denoising.* In: *IEEE Transactions on Image Processing* 17.12 (2008), pp. 2324–2333.

[83] **Zhou, J. and Gu, J.** *A model–based method for the computation of fingerprints' orientation field.* In: *IEEE Transactions on Image Processing* 13.6 (2004), pp. 821–835.

[84] **Zhou, J., Xin, L., and Zhang, D.** *Scale-orientation histogram for texture image retrieval.* In: *Pattern Recognition* 36 (2003), pp. 1061–1063.

**Forschungsberichte
aus der Industriellen Informationstechnik
(ISSN 2190-6629)**

Institut für Industrielle Informationstechnik
Karlsruher Institut für Technologie (KIT)

Hrsg.: Prof. Dr.-Ing. Fernando Puente León, Prof. Dr.-Ing. habil. Klaus Dostert

Die Bände sind unter www.ksp.kit.edu als PDF frei verfügbar oder als Druckausgabe bestellbar.

Band 1 Pérez Grassi, Ana
**Variable illumination and invariant features for detecting
and classifying varnish defects.** (2010)
ISBN 978-3-86644-537-6

Band 2 Christ, Konrad
**Kalibrierung von Magnet-Injektoren für Benzin-
Direkteinspritzsysteme mittels Körperschall.** (2011)
ISBN 978-3-86644-718-9

Band 3 Sandmair, Andreas
**Konzepte zur Trennung von Sprachsignalen in
unterbestimmten Szenarien.** (2011)
ISBN 978-3-86644-744-8

Band 4 Bauer, Michael
**Vergleich von Mehrträger-Übertragungsverfahren und
Entwurfskriterien für neuartige Powerline-Kommunikationssysteme
zur Realisierung von Smart Grids.** (2012)
ISBN 978-3-86644-779-0

Band 5 Kruse, Marco
**Mehrobjekt-Zustandsschätzung mit verteilten Sensorträgern
am Beispiel der Umfeldwahrnehmung im Straßenverkehr** (2013)
ISBN 978-3-86644-982-4

Band 6 Dudeck, Sven
**Kamerabasierte In-situ-Überwachung gepulster
Laserschweißprozesse** (2013)
ISBN 978-3-7315-0019-3

Band 7 Liu, Wenqing
**Emulation of Narrowband Powerline Data Transmission Channels
and Evaluation of PLC Systems** (2013)
ISBN 978-3-7315-0071-1

**Forschungsberichte aus der Industriellen Informationstechnik (ISSN 2190-6629)**

Institut für Industrielle Informationstechnik | Karlsruher Institut für Technologie (KIT)


Hrsg.:     Prof. Dr.-Ing. Fernando Puente León, Prof. Dr.-Ing. habil. Klaus Dostert

Band 8     Otto, Carola
           **Fusion of Data from Heterogeneous Sensors with Distributed
           Fields of View and Situation Evaluation for Advanced Driver
           Assistance Systems.** (2013)
           ISBN 978-3-7315-0073-5

Band 9     Wang, Limeng
           **Image Analysis and Evaluation of Cylinder Bore Surfaces
           in Micrographs.** (2014)
           ISBN 978-3-7315-0239-5